Plankton and Fisheries

John Grahame
Ph.D.
Lecturer in Zoology and Ecology
University of Leeds

Consultant Advisor: Professor M.A. Sleigh

Edward Arnold

First published in Great Britain 1987 by
Edward Arnold (Publishers) Ltd, 41 Bedford Square, London WC1B 3DQ

Edward Arnold (Australia) Pty Ltd, 80 Waverley Road, Caulfield East,
   Victoria 3145, Australia

Edward Arnold, 3 East Read Street, Baltimore, Maryland 21202, U.S.A.

British Library Cataloguing in Publication Data

Grahame, John
   Plankton and fisheries.
   1. Marine plankton
   I. Title
   574.92   QH91.8.P5

   ISBN 0-7131-2941-7

Text set in 10/11pt Times
by Colset Private Limited, Singapore
Made and printed in Great Britain
by Richard Clay plc, Bungay, Suffolk

Preface

This book is addressed principally to undergraduate students. I hope it will be of interest to those studying marine biology, and also to those who are not but who would like to know something about the study of plankton and fisheries.

Plankton ecology has a central place in marine biology. The plankton has attracted attention because of its fascination, and also because of the essential role of plankton in marine food chain dynamics and fish production. An understanding of plankton ecology demands knowledge of currents and the structure of the water column, of physical factors such as light and nutrients, and of a variety of biological factors, some physiological and others from field ecology. To understand how food energy is transferred through the food web we need an understanding of energy flow and the properties of food webs. An important starting point in fisheries biology is the population dynamics of fish and the responses of their populations as prey under predation from man.

My aim is to explore some of the facts and ideas relating to these topics, making reference, where necessary, to the general ecological literature outside marine biology. Too often it seems as if workers in one sphere neglect the findings that come from others and I have deliberately chosen a wide canvas. This has inevitably meant that material has been treated in a highly selective way – this is not a comprehensive text on the plankton, nor on fisheries, nor on oceanography. Rather, I have tried to achieve a synthesis drawing on these fields which will be accessible to those who wish to know how they are related, and the importance of this area to man. If subsequently the reader wishes to specialise, I have included an extensive bibliography with which he may make a start.

I am grateful to Professor R.McN. Alexander, Dr J.F. Allen, Dr E.S. Grahame, and Dr P.J. Mill for reading parts of the text and making useful criticisms. Professor M.A. Sleigh has been a helpful and encouraging reader and I am thankful for his support.

Leeds, 1987   John Grahame

For Suzanne

Contents

1

Circulation in the oceans

The flow of water in ocean surface currents is of crucial importance to the organisms living in these habitats. Not only do the currents affect their distribution, but also the processes associated with current flow may play a major role in controlling levels of production. Therefore, an understanding of some of the surface current processes is necessary for our purposes. Deep currents are important in general oceanography, but since they are not so intimately associated with plankton and production, they are given no treatment here.

The impelling force for the surface circulation comes from atmospheric winds, which in turn are driven by differential heating of the earth by the sun. On a large scale, the major features of atmospheric, and oceanic, circulation are quite constant. In the atmosphere, pressure is relatively high at mid-latitudes

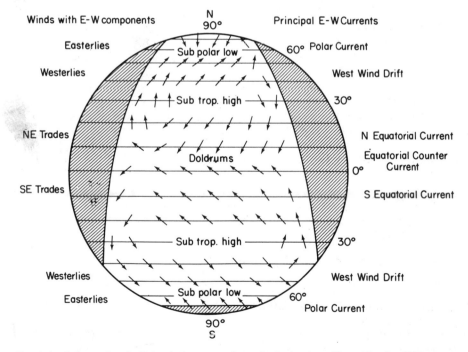

Fig. 1.1 Schematic wind circulation over a hypothetical ocean. (From Fleming R.H. (1957). In: *Treatise on marine ecology and paleoecology, Vol. 1, Ecology*. J.W. Hedgpeth (ed). Geological Society of America Memoir, 67.)

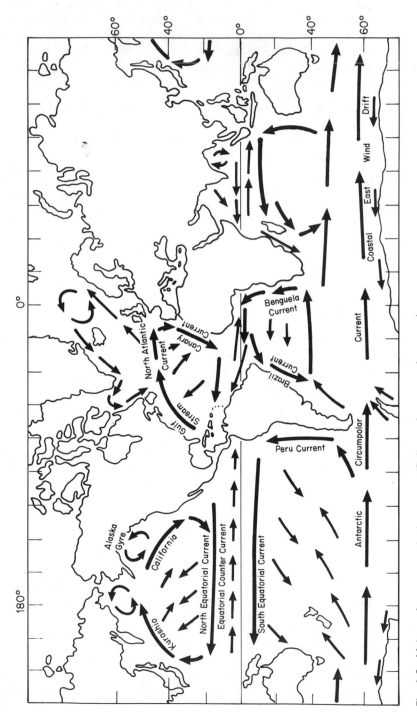

Fig. 1.2 Major surface currents in the oceans. (Redrawn from several sources.)

over the eastern parts of the oceans. This results in the occurrence of mid-latitude anticyclones, sometimes called the subtropical anticyclones. These anti-cyclones are associated with belts of winds: easterly near the equator and westerly at about 40° north or south. It is conventional to express wind direction in terms of where the wind comes from, thus the equatorial winds blow from the east (and towards the equator) as the North- or Southeast Trades. The 'roaring forties' winds blow from the west (and towards the poles). An idealised representation of the wind system is shown in Fig. 1.1.

These winds excite movement in the surface waters through frictional drag. It is held that the major driving force for the surface circulation of the oceans comes from the low latitude trade winds. In the major ocean basins there is thus a circular flow of water round the basin, clockwise in the northern hemisphere and counterclockwise in the southern (Fig. 1.2). These circular systems are called gyres, and they are a feature of the North and South Atlantic and Pacific Oceans, and of the South Indian Ocean. It is conventional to express current direction in terms of where the current is flowing towards, thus the equatorial currents are westward and those on the opposite sides of the gyres eastward. The gyres must not be seen as driving one another: they are independent comple-mentary systems in the hemispheres, each associated with the appropriate major atmospheric features.

Circulation in the Atlantic

The Atlantic Ocean is relatively well known and it is useful to start with this example. Across the equatorial region flow the North and South Equatorial Currents, carrying water from east to west (Fig. 1.3). Some of the South Equa-torial Current water crosses into the North Atlantic, the rest turns south as the Brazil Current flowing down the South American coast. The Brazil Current turns east and flows as part of the Antarctic Circumpolar Current, a branch of which turns north along the west coast of Africa as the Benguela Current. In turn, this gives rise to the South Equatorial Current.

The Equatorial Currents tend to pile water up on the western side of the ocean basin. In between them is the Equatorial Counter Current, flowing east between 9° and 5°N. This current is often absent from the western surface region in the first six months of the year, while in the eastern region it is a more permanent surface feature. It tends to flow where there is least wind stress – i.e. under the equatorial doldrums. Also flowing east, but at 60–100 metres depth, and on the equator, is the Equatorial Undercurrent (not represented in Fig. 1.3). Again, this is more reliably found in the eastern portion of the Ocean.

The North Equatorial Current is joined on the west side of the Ocean by a component of South Equatorial Current water, part of this combined flow passes outside the Antilles island arc as the Antilles Current, the rest flows through the Caribbean Sea. From here water escapes into the Atlantic as the Florida Current, passing between Florida and Cuba. It is joined by the Antilles Current and where the flow breaks away from the North American coast, at about Cape Hatteras, the current is called the Gulf Stream. The Florida and Antilles Currents can be distinguished by slightly different salinities, reflecting their origins. The Florida Current draws water from the north branch of the South Equatorial Current, which flows into the Caribbean after crossing the

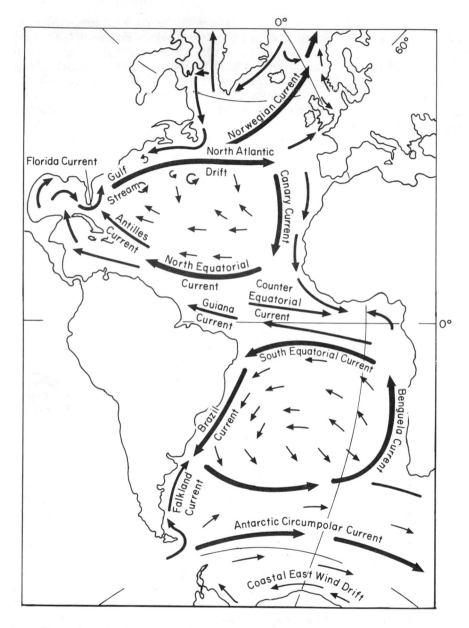

Fig. 1.3 Surface circulation in the Atlantic Ocean. (Redrawn from several sources.)

equator. This water has a component of Antarctic Intermediate water, which is formed at about 50°S in the Southern Ocean with a salinity of about 34.2‰. It flows north through the Atlantic at about 1000 metres depth, mixing with higher salinity surface water and lowering the salinity of this water. Some of this gets entrained in the South Equatorial Current, and this lowers the salinity of the

Florida Current very slightly when compared with that of the Antilles Current, the latter having been derived entirely from North Equatorial Current Water.

The Gulf Stream flows northeast and across the Atlantic from the Grand Banks of Newfoundland, when it is called the North Atlantic Current. On the western side of the Ocean this divides, some of the water flowing north between Scotland and Iceland, the rest flowing south as a broad slow current over the eastern portion of the Ocean.

This is a simple version of the classical account of Atlantic circulation. For more detail the reader is referred to Pickard and Emery (1982).

Circulation in the Pacific

The general features of circulation in the Pacific are shown in Fig. 1.2. The equatorial system is particularly well developed, and has been well studied. As in the Atlantic, there are westward North and South Equatorial Currents, with an Equatorial Counter Current between them. The eastward undercurrent is called the Cromwell Current. The situation is complicated by the existence of two further currents, both comparatively weak, south of the equator. The South Equatorial Counter Current flows eastward south of the South Equatorial Current, while further south still there is another westward current. From the North Equatorial Current there flows the Kuroshio Current leading to the North Pacific Current. This divides in the eastern north Pacific to give the California Current flowing south and the Alaskan Gyre in the Gulf of Alaska. The California Current returns water to the North Equatorial Current.

It is apparent that in most respects the North Pacific is the analogue of the North Atlantic. Circulation in the South Pacific is relatively poorly understood, but the main features of the oceanic gyral system are observed.

Circulation in the Indian Ocean

The Indian Ocean is unusual when compared with the Atlantic or Pacific in that it is truncated in its northern extent by the land masses of Asia. Because of this land there is a seasonal change in the winds north of the equator. Between November and March the winds blow over the equatorial Ocean as they do elsewhere, north of the equator they are here called the Northeast Monsoon instead of the Northeast Trades. From May to September the northern system reverses and the Southeast Trades (which continue throughout the year) extend across the equator (shifting in direction as they do) and blow across the northern Indian Ocean as the Southwest Monsoon. Thus while the Indian Ocean south of the equator has a typical gyre, and during the Northeast Monsoon there is a North Equatorial Current, during the Southwest Monsoon circulation in the northern Indian Ocean largely reverses and the westward North Equatorial Current is replaced by an eastward Southwest Monsoon Current, flowing with the Equatorial Counter Current. During the Northeast Monsoon an equatorial undercurrent is evident, during the Southwest Monsoon it is lost in the general eastward flow at the equator.

Circulation in the Southern Ocean

The Southern Ocean is uninterrupted by land barriers, and the ocean currents run around the globe – the only place in which they do. There is a narrow coastal current flowing westward, the East Wind Drift. The rest of the Southern Ocean is dominated by the eastward flowing Antarctic Circumpolar Current (or West Wind Drift). It is necessary here to recall the reversed conventions used in naming winds and currents, explained above. The account given below is particularly for the relatively well understood Atlantic sector of the Southern Ocean.

Within the Antarctic Circumpolar Current there are more slow but very significant flows of water south towards the continent and north away from it, these flows occurring at different depths (Fig. 1.4). On the surface, there are two convergence zones which are used to delimit the boundaries of the Southern Ocean. At about 50°S in the Atlantic, and 60°S in the Pacific Ocean sector, is found the Antarctic Polar Front (Antarctic Convergence). The surface temperature rises abruptly from about 4° to 8°C. At about 40°S is the Subtropical Convergence, where there is a further rapid rise of temperature by about 4°C. It is usual to regard the region between the continent and the Antarctic Polar Front as the Antarctic zone, while that between the two convergences is referred to as the Subantarctic zone. The reasons for the occurrence of the convergences are not well understood, and indeed recent work has shown that the picture is more complex than the account given above indicates, particularly with respect to the Antarctic Polar Front. However it is well established that there are north-south and vertical movements of water as shown in Fig. 1.4. Deep water of relatively high salinity flows south and upwells off the continent, associated with a divergence between the eastward and westward surface currents. Some of this water flows south in the region of the Weddell Sea where it is further cooled and may

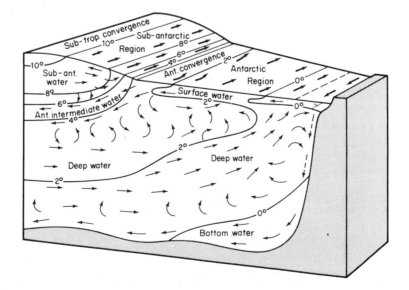

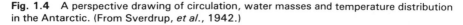

Fig. 1.4   A perspective drawing of circulation, water masses and temperature distribution in the Antarctic. (From Sverdrup, *et al.*, 1942.)

be made more saline by freezing of sea ice. This helps to form a very dense water which sinks to form Antarctic Bottom water and which then flows north along the deep ocean floor.

Water also flows north from the offshore divergence along the surface. This water is diluted by precipitation, and by the time the flow approaches the Antarctic Polar Front it has a salinity of 33.8‰. While it floats over the deep water which is of similar temperature but higher salinity, when it meets the warmer Subantarctic Surface water it sinks beneath it forming Antarctic Intermediate water. This water mass is of interest in that it can be traced for great distances north in the Atlantic; as we have seen its influence can be detected in the South Equatorial Current, and in the Florida Current in the northern hemisphere.

Coriolis force and its consequences

According to Newton's laws of motion, a body will continue in a state of uniform motion unless it is acted on by a force, when its velocity will be changed. Velocity is a vector quantity, which implies that it has direction as well as speed. The uniform direction expected under Newtonian laws is apparent with respect to fixed rectangular coordinates. Thus, if a marble is set rolling across a sheet of graph paper resting on an absolutely flat surface, it will roll in a straight line with respect to the lines on the paper. The negative acceleration which ultimately will halt it is due to frictional forces.

Large-scale motion across the earth's surface is complicated by the fact that

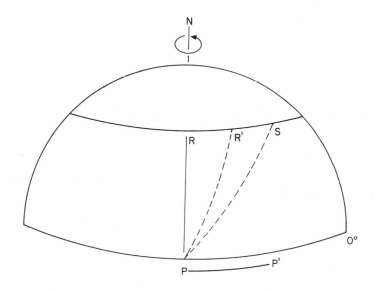

Fig. 1.5 The paths of a frictionally bound body and a free particle across the earth when both begin moving towards the north pole from a position on the equator. The bound body will start from P towards R, velocity P R. As the earth is rotating eastward and the speed of rotation decreases with increasing latitude, the bound body loses eastward speed (velocity P P' decreases) and will reach the latitude of R and R'. The free particle will also start from P towards R, but retaining all its initial eastward speed (velocity P P' maintained) it will arrive at the latitude of R at S.

the familiar coordinates are not fixed, but rotating with the earth and, more-over, the x and y coordinates (longitude and latitude) are curved with the surface of the globe. Imagine a car at the intersection of a meridian and the equator. While appearing stationary to a terrestrial observer, the car nevertheless has an eastward velocity due to the earth's rotation – although this would be apparent only to an observer in fixed coordinates independent of terrestrial rotation.

If the car is now driven due north (Fig. 1.5), it will be tightly frictionally bound to the earth's surface. If a projectile is fired in the same initial direction as the car, it will be above the surface, not frictionally bound to it. Supposing that the projectile suffers no air resistance, it will have the same initial eastward velo-city as the car. As both proceed north the paths of the projectile and car will diverge. This will happen because the car will lose eastward velocity as it travels north, while the projectile will retain its initial eastward velocity. This diver-gence is due to the Coriolis effect, and to explain it an apparent force, Coriolis force (CorF) is introduced. The CorF is said to act on the particle causing it to deviate from the initial velocity when this is defined in terms of terrestrial coor-dinates. In fact, of course, in the example given above, the projectile is more nearly obeying Newton's laws of motion in terms of fixed rectangular coor-dinates than is the car. Notice that in the above example motion was in the northern hemisphere and the deflection of the projectile was to the right. In the southern hemisphere the same considerations apply, but deflection would be to the left.

It is less obvious why there should be a similar deflection in the case of motion along a line of latitude (other than the equator). Imagine a plane surface tan-gential to the earth's surface at some latitude (Fig. 1.6). This plane represents a section through fixed rectangular coordinates, and a moving body would be expected to travel straight across it. However, the surface of the earth under the

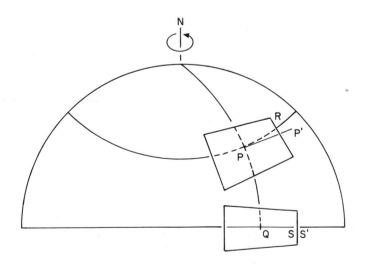

Fig. 1.6 Relative motion on the curved surface of the earth and on imaginary flat plane surfaces, at the equator and at 45°N. The apparent deflection is not attributable to Coriolis force as such, but rather to centrifugal force. It operates to have the same outcome – apparent deflection *cum sole*.

plane surface is curved. A projectile fired due east from point P will travel along the straight line P P′ across the plane, while a car proceeding in the same direction will travel along the line P R, following the curvature of the earth. Once again in the northern hemisphere there is deflection to the right, in the southern hemisphere the same phenomena cause deflection to the left. To an observer in the northern hemisphere, facing the equator, the sun appears to move to the right across the sky, in the southern hemisphere the sun appears to move to the left. Therefore this deflection is said to be 'with the sun' – *cum sole*. For movement due east along the equator, from point Q to point S in Fig. 1.6, the deflection out of the plane surface carries the car straight down – i.e. while there is a vertical component there is no horizontal component and point S on the earth's surface lies directly beneath S′ on the plane surface.

The deflection due to CorF (Coriolis force) is apparent for bodies not tightly frictionally bound to the earth. Water and air masses are in this category, and both show deflection when they move as currents or winds. The effects are seen in the familiar patterns of winds in cyclones (or anticyclones). However, such deflections would not be expected to be seen in bodies which are tightly frictionally bound to the earth. For example in the case of the car, CorF would be of the order of one thousandth of other forces at latitude 40° (Sverdrup *et al.*, 1942).

The magnitude of CorF for any given body under a set of circumstances can be calculated using the Coriolis parameter, f.

$$f = 2 \, \Omega \sin \phi$$

where Ω = angular velocity of the earth in radians per second
           ϕ = latitude.

CorF is now mass × velocity × f, or $m \, f \, u$.

For a series of bodies of the same mass, CorF will therefore depend on latitude (ϕ) and velocity, (u), increasing with both. Although CorF increases with velocity, for any value of ϕ it remains a constant proportion of u, and the relative deflection of fast-moving bodies is much smaller than that of slow-moving bodies. This can be seen with two examples. Consider the cases of the Florida and Canary Currents, both of which pass 30°N. Take the example where $\phi = 30°$, $\sin \phi = 0.5$. The speeds of the currents may be taken as 200 cm s⁻¹ (Florida) and 25 cm s⁻¹ (Canary). Expressing CorF in terms of an acceleration (i.e. force/mass) and using unit mass, the time taken to travel 100 m, and the deflection occurring in that time can be calculated, obtaining the following values:

	u (cm s⁻¹)	CorF (cm s⁻¹)	Time to travel 100 m (seconds)	Deflection (m)
Florida Current	200	0.0146	500	18.25
Canary Current	25	0.00183	4000	146.4

The deflection is calculated using the expression

$$\text{deflection} = \tfrac{1}{2} \, a \, t^2$$

where a is the acceleration due to CorF and t is time.

Note that while the ratio CorF/*u* is the same for both examples, the relative deflection is much greater for the slow Canary Current than it is for the fast Florida Current.

The fact that slowly moving bodies suffer greater deflection from CorF than do faster moving ones, leads to a complex structure in wind driven surface currents. A wind blowing over the sea will excite a current in the surface layer. This current will be slower than the wind and show greater deflection *cum sole* than does the wind. The surface current will in turn excite a slower flow beneath it, and this will be deflected still more. This process continues down through the topmost layers until the currents are very slow and flowing in an opposite direction to the wind. This structure is known as an Ekman spiral.

During the cruise of the *Fram* in the Arctic (1893–1896) the Norwegian oceanographer F. Nansen noticed that the ice drift deviated to the right of the wind direction. He attributed this to an effect of the earth's rotation, and supposed that there would be a series of successive layers of flow deviating in the same way. At Nansen's suggestion, Ekman undertook the mathematical analysis which confirmed these suppositions.

From Ekman's equations, it can be predicted that the surface current will be between 1% and 3% of the speed of the wind driving it, and will deviate from the wind direction by 45° *cum sole*. The predicted net transport will be at 90° to the wind direction – this net transport is the so-called Ekman transport.

The conditions for development of an Ekman spiral – steady wind, infinite and homogeneous ocean, and no other forces acting – are not met in practice, and a full Ekman spiral has not been observed in the oceans. However, Ekman's analysis does serve to explain important features of surface circulation, and, in particular, the phenomenon of upwelling, with which we shall be much concerned. In this connection, it is interesting to note that one of Ekman's assumptions, that of a steady wind blowing for a long time, is well approximated in the Trade wind regions about the equator.

Western boundary intensification

In the Atlantic, Indian and North Pacific oceans, there is a phenomenon known as western intensification. The currents flowing polewards on the western sides of the oceans are swift and concentrated when compared with those flowing towards the equator on the eastern sides, which are slow and diffuse. Such western intensification seems not to occur in the western South Pacific. This phenomenon is discussed in terms of what is called vorticity. If an object rotates counterclockwise, it is said to have positive vorticity, if clockwise, negative vorticity. For example, in the North Atlantic, remembering that the whole flow is clockwise and wind driven, the wind is imparting negative vorticity to the ocean over the system as a whole. As the currents flow north along the western boundary, negative vorticity is again increasing. This is because of the increasing value of the Coriolis parameter, f, reflecting the increasing tendency for the water to move *cum sole* – to the right, or clockwise. Thus there is an overall increase in negative vorticity. In the eastern boundary, the water flowing south experiences increasing negative vorticity from the wind, but decreasing negative vorticity (i.e. positive vorticity) from the decreasing value of f, and the two cancel. If ζ_p represents planetary vorticity (due to f) and ζ_τ wind stress vorticity

eastern boundary current: $+ \zeta_p - \zeta_\tau \approx 0$
western boundary current: $- \zeta_p - \zeta_\tau \neq 0$

and the system as a whole is out of balance. In order to restore
strong current shear in the western, but not in the eastern, boun
needs to be postulated. In the example given here, current shear in the Florida
Current and Gulf Stream arm of the gyre imparts positive vorticity ($+ \zeta_s$) so that

$$- \zeta_p - \zeta_\tau + \zeta_s \approx 0$$

The current shear is evident as a considerable slowing of the current speed from
a maximum value offshore to lower speeds as the coast is approached, in the
western boundaries of the gyre systems.

Similar considerations apply in other oceans, though, as has been mentioned,
intensification is not evident in the southwestern Pacific.

Upwellings

Some of the processes in surface currents, which are of great importance to the
organisms living there, can now be understood. Currents not only move water
horizontally, but may cause it to move vertically also – it has already been seen

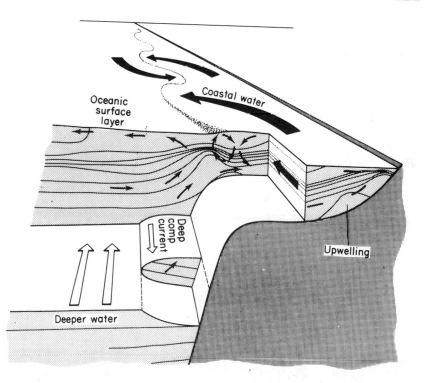

Fig. 1.7 A perspective drawing showing an idealized picture of water movements in an
upwelling, based on the Benguela Current upwelling. (From Hart and Currie, 1960).

that the so-called Deep Water in the Southern Ocean upwells at the divergence off the Antarctic continental coast. Upwellings of one sort or another are of widespread occurrence. For upwelling to be of importance to planktonic production, it must occur over a period of time (weeks at least) and such upwellings are found in the slow, diffuse eastern boundary currents, in the equatorial currents, and around the Antarctic continent. Eastern boundary currents will be considered first.

It often happens that the wind blows for some time towards the equator over these currents, more or less parallel to the nearby continental shore. Ekman transport then moves surface water away from the coast, and this water is replaced by water upwelling from below, usually from depths of about 50 to 300 m. A good example is the structure inferred for the Benguela Current, and believed to be representative of coastal upwellings in general. In this case, the upwelling zone was found to be some 65 to 130 km wide and 1600 km long. A width of 130 km may be exceptional; another value is, for example, about 10 km in the case of the California Current. Figure 1.7 shows the movement of surface waters away from the coast and its replacement by deeper water. Vertical movement was typically of the order of 1 to 5 m per day, though locally it might exceed this for short periods. Offshore, where the cool upwelled water meets the warmer oceanic water, there was a convergence/divergence system, with a secondary upwelling beyond this. Underneath the upwelling was a polewards compensation current, which seems to be related to the equatorwards transport in the surface layer.

Although the upwelled water comes from no very great depth, it tends to be relatively rich in the inorganic nutrient ions needed for phytoplankton growth. For example, Hart and Currie (1960) reported up to 2 μg atoms of phosphorous (as phosphate) per litre in the surface waters of the Benguela Current during upwelling, with levels declining to zero in the open ocean. Figure 1.8 shows the distribution of phosphate in a section off the mouth of the Orange river, in west southern Africa, in September, during intense upwelling. The upward slope of

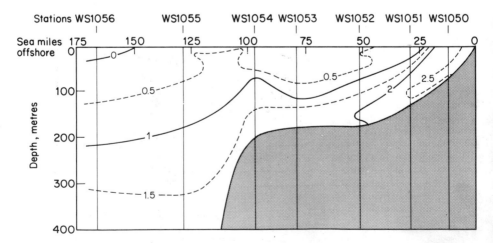

Fig. 1.8 The distribution of phosphate as mg atoms P per cubic metre, in the Benguela Current upwelling in September. (From Hart and Currie, 1960.)

the isopleths of phosphate to the surface near the shore is indicative of the upwelling. Between 160 and 200 km offshore the isopleths rise again, in the region of the secondary upwelling beyond the convergence zone. It is from data such as these, as well as from current measurements, that oceanographers deduce the direction of current flow. It must be realised that phosphate is what is termed a non-conservative constituent of sea water: it undergoes change as a result of processes in the water column, in this case as a result of uptake by organisms. This is reflected in the failure of phosphate levels at the surface to reach those of deeper water over any great area – as phosphate is upwelled, it is used. The reason for the high values over the continental shelf is the regeneration of inorganic phosphate from sinking organic matter descending from the surface.

The importance of the supply of nutrients from subsurface water in enhancing algal production has been recognised since the first studies of upwelling. Where upwelling brings to the surface nutrient-poor water, as happens in the Atlantic off the Brazilian coast and may happen also in the Peru Current, there is no enhanced productivity. Nonetheless, the usual primacy given to nutrient levels as such has been challenged (see chapter 4).

Another feature of coastal upwelling systems is that they may act as nutrient traps. Nutrients are transported into the system in the lateral flow before upwelling. They are upwelled and incorporated into living organisms, thence into

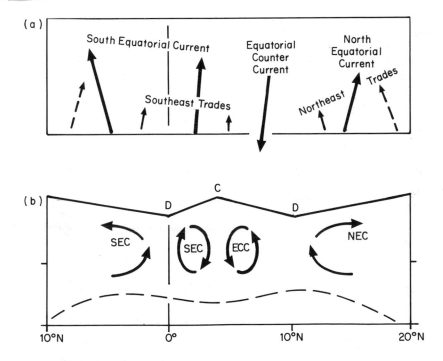

Fig. 1.9 Schematic representations of equatorial circulation. (a) Winds (narrow arrows) and surface currents (thicker arrows) at the equator, in plan view. (b) Surface currents in sectional view. The slope of the sea surface is shown much exaggerated. The broken line indicates the permanent thermocline. D, divergence; C, convergence. (Redrawn from Pickard and Emery, 1982, and other sources.)

faeces and corpses which sink and may re-enter the inward flow, thus becoming to some extent trapped in the system. Regeneration of nutrients from organic matter reaching the bottom over the continental shelf will add to this process.

Equatorial Current upwelling

Upwelling is also characteristic of the equatorial current systems, where it is most intense in the eastern parts of the oceans. In both the Atlantic and Pacific Oceans the surface Counter Current lies a few degrees north of the equator. The westward South Equatorial Current straddles the equator, and the reversal of sign of CorF across this flow sets up a divergence in the current (Fig. 1.9) with Ekman transport away from the equator in both hemispheres. Water must upwell along the equator to balance the flow away from the divergence. Between the northern extent of the South Equatorial Current and the Counter Current, Ekman transport leads to a convergence and downwelling on one side of the Counter Current and a divergence and upwelling on the other side, where it bounds on the North Equatorial Current. These convergences and divergences are reinforced by variations in the strength of the winds. Wind speeds are lowest in the doldrums over the Counter Current. As wind speeds increase into the Trade Wind zones over the Equatorial Currents, Ekman transport increases. This reinforces the divergences already established by the change in sign of CorF across the equator and the difference in direction between the Equatorial Currents and the Counter Current.

Again, these upwellings are responsible for bringing to the surface inorganic nutrient ions, as can be seen from the distribution of phosphate and silicate in the equatorial Pacific Ocean (Fig. 1.10).

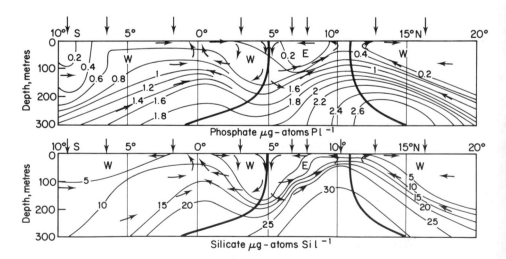

Fig. 1.10 Phosphate and silicate in a vertical section in the Pacific Ocean between 10°S and 20°N in the Pacific Ocean. The data were obtained during the cruise of the *Carnegie* in 1928–29. (From Sverdrup 1942.)

Stratification and seasonal overturn

It has been pointed out that surface currents flow in the top few hundred metres of the oceans, and that upwellings do not bring water to the surface from more than about 300 m, except in the Southern Ocean. The greater volume of the ocean water is cold, dark and very slow-moving. All of the significant heat inflow to the ocean occurs at the surface – heat flux from the earth's crust is trivial in this context. Nearly all the infra-red radiation from the sun reaching the sea is absorbed in the first metre of water, heating the extreme surface layer. From this layer, heat can be transported downwards very slowly by conduction or much more rapidly by turbulent mixing processes. This turbulent mixing will have to work against the tendency for a stable, stratified water column to be established whenever surface heating is sufficient to crea: ; a warm, buoyant surface layer. The upper portion of the water column through which temperature is fairly constant with depth is known as the mixed layer, in shallow seas it may extend to the bottom but in deeper water it overlies colder, denser water. The mixed layer will be of varying thickness, depending upon the intensity of surface heating and the vigour of mixing processes.

   In temperate latitudes, summer warming causes the development of a warm surface layer which is clearly separated from the water underneath. The zone of

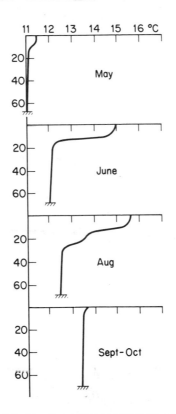

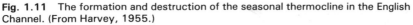

Fig. 1.11   The formation and destruction of the seasonal thermocline in the English Channel. (From Harvey, 1955.)

thermocline ✳

rapid change of temperature with depth is known as the thermocline, in this case a seasonal thermocline. It is associated with an equally dramatic change in density, so that the thermocline coincides with what is called a pycnocline. This represents a considerable barrier to movement of water across it, and the greater the density discontinuity, the more effective is the isolation of the two layers of water with respect to vertical movement of water and salts.

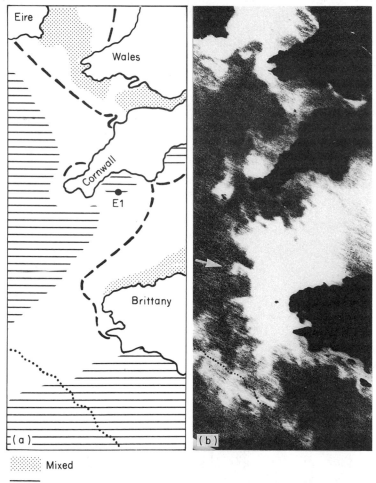

:::: Mixed

≡≡ Stratified

Fig. 1.12 **(a)** The positions of fronts (broken lines) in the southern Irish Sea and western English Channel, predicted by a numerical model of tidal mixing. The fronts lie in a band of transitional water, between stratified water (horizontal lines) and mixed water (stipple). The position of the station E1 is indicated by a solid circle. **(b)** Satellite photograph of the same area for August 20, 1976, showing cool mixed water (white) and warm stratified water (dark). The English Channel front lies somewhat to the west of the predicted position. It shows an irregular structure, and the hook-like feature (arrow) may be interpreted as evidence of eddy formation. In both figures the black dotted line indicates the 200 m depth contour. (From Pingree, 1978.) Satellite image by permission of The University of Dundee.

In autumn, the surface layers begin to cool, and wind mixing tends to make the water column homogeneous with respect to temperature. This lasts until the onset of stratification the following spring. This process has often been illustrated using graphs showing temperature distribution with depth from the station 'E1' in the English Channel (Fig. 1.11).

The classical understanding of summer stratification has recently been modified for very shallow shelf seas such as the English Channel and North Sea. Here, tidal mixing is often sufficient to maintain a mixed water column through spring and summer, resulting in relatively cool surface water. As mixing is continuous, surface heat is dispersed downwards, and the surface layers remain relatively cool and nutrient-rich. There is a tendency for fronts – abrupt changes in temperature and density on a horizontal scale – to form between this very shallow, mixed, cool water, and the somewhat deeper, stratified water further offshore. Eddies may now form along the fronts, leading to the transfer of nutrients across from the cool mixed water column to the stratified water column. This injection of nutrients boosts phytoplankton production in the surface of the stratified column.

Using predictions from tidal models and satellite photographs, Pingree (1978) demonstrated a good relationship between the predicted and actual occurrence of fronts in waters around the British Isles (Fig. 1.12). In view of the proximity of E1 to the western English Channel front, and the predicted existence of well-mixed summer water in this region (Fig. 1.12), it might be wondered how the

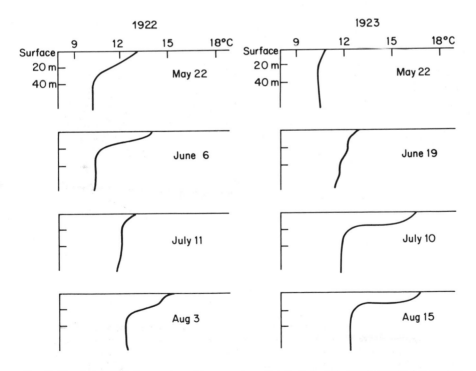

Fig. 1.13 A detailed time series of temperature distribution with depth graphs for station E1 in the English Channel. (From Harvey, 1925.)

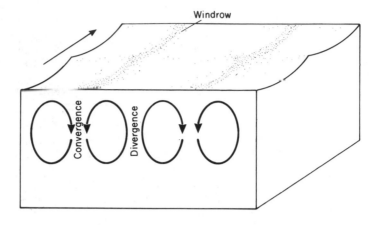

Fig. 1.14   Schematic drawing of Langmuir circulation cells. A 'windrow' is shown over the convergences, this is absent over divergences.

smooth sequence of changes implied by Fig. 1.11 comes about. In fact the detailed work of Harvey (1925) shows marked changes in stratification between months in a year and also between years (Fig. 1.13). Harvey (1925) noted that there was evidence of water movement past E1 during 1922–1923, although this was a relatively stable period compared with 1921 when there was a major incursion of Atlantic water. Certainly Fig. 1.13 is more consistent with what is now known of the western English Channel than is Fig. 1.11.

Surface mixing maintains a mixed layer in the top of the water column, and as we have seen this may lie over a distinct thermocline. Mixing is also of considerable importance in keeping phytoplankton cells in suspension (Chapter 3). An important mechanism here is Langmuir circulation.

Langmuir (1938) described observations he made from a ship in the north Atlantic on 7 August, 1927. Floating seaweed occurred in 'streaks' up to 500 m long, and between 100 to 200 m apart. There were smaller streaks between the main large ones. All were lined up parallel with the wind direction, and when the next day the wind abruptly changed direction through about 90°, within about 20 minutes all the streaks were realigned parallel to the new wind direction. Langmuir deduced that the wind must be causing alternating right and left helical vortices in the water, with the result that water was rising between the streaks to a divergence and flowing across the surface. Where two streams met a convergence was formed, and the water sank. Floating material would gather along the lines of the convergences, hence the formation of the streaks. Over subsequent years Langmuir made a fascinating series of experiments on Lake George, New York State, using oils, dyes, floating leaves, cork buoyed string and other apparatus, which confirmed his deductions about helical vortices. These have come to be known as Langmuir circulations, and are illustrated in Fig. 1.14. As Langmuir noted, while streaks or windrows are easily and frequently observed on lakes and at sea they had received no attention from oceanographers, and indeed this remains almost true today (Faller and Caponi, 1978). Windrows may have other causes, but Langmuir circulations are often involved. Windrows on Malham Tarn, Yorkshire, are shown in Fig. 1.14.

Fig. 1.15 Windrows on Malham Tarn, North Yorkshire, June 1985. The presence of lines of foam and debris roughly in parallel with the wind is evidence of Langmuir circulation in the water underneath.

If the convergence downwelling is strong enough, heat will be carried down in warm water from the surface. Scott *et al.* (1969) have demonstrated that this happens in lakes, confirming Langmuir's (1938) suggestion that this is a process contributing to the formation of a mixed layer overlying a summer thermocline in lakes. There is no reason why this should not be true in the sea as well. It is likely that more than one mechanism is involved in producing the helical vortices, but these remain obscure. Faller (1969) suggested that the exponential decay of the vertical oscillations due to waves, which are maximal near the free (upper) surface and diminish downwards, might give rise to instability. He referred to this as the 'eddy pressure concept'. The instability would be analogous to thermal convective instability, first studied by Benard (Defant, 1961). In this phenomenon, if a thin surface layer of liquid is cooled, causing it to become more dense and to sink, the whole mass of the liquid divides into convection cells of the same sort as Langmuir circulations. Defant (1961) ascribed the formation of lines of foam on the sea surface to Benard convection cells. However the observations of Langmuir (1938) and Scott *et al.* (1969) show wind to be important as well.

More recently it has been shown that an interaction between wind and waves is necessary for Langmuir circulations, and this distinguishes them from Benard cells. Faller and Caponi (1978) stated that the 'eddy pressure concept' is simplistic and incorrect, preferring wind-wave interaction mechanisms. Understanding of the models of Langmuir circulations involves an excursion into fluid mechanics which is beyond the scope of this book; the interested reader is referred to the papers cited. Whatever the mechanism, such circulations do occur, as has been shown by observations in the field.

The scale of Langmuir circulations is very small in terms of the whole ocean. Observations on Lake George suggested that the helical vortices might extend no further than 10 or 15 m below the surface in summer when there was a thermocline (Langmuir, 1938). However in spring and autumn when the water column was isothermal, they might extend to the bottom (about 50 m), although they were said to be diffuse at depth. The spacing between windrows is an indicator of the size of the circulation. Assaf *et al.* (1971) showed that in strong winds (5 to 15 m s^{-1}) there develops a hierarchy of Langmuir circulations, with spacing between convergences of the order of 280 m, 35 m and 5 m. They considered that Langmuir circulations are the most common and effective mechanism of vertical transport in the mixed layer of the seas.

Vertical current velocities were estimated to be about 1.6 cm s^{-1} (Langmuir, 1938) in downwellings. Values between ~3 to ~7 cm s^{-1} were reported by Scott *et al.* (1969) for lake and oceanic conditions, downwelling speed being directly proportional to wind speed between 3 and 7 m s^{-1}.

Circulation in the oceans, from the scale of the major currents (many hundreds of kilometres), to frontal eddies (a few tens of kilometres in diameter) and Langmuir circulations (tens to hundreds of metres) operates to transport water, materials, organisms and heat in the oceans. These processes are of profound importance to the plankton. For an understanding of why this is so, something of the distribution of nutrients and light must be known.

2

Light and nutrients

Light in the sea

The availability of light

Light is a flux of radiant energy, or of photons or quanta, for light has both wave and particulate properties. It may be measured in terms of irradiance, with units in energy as watts (W) or in particles as quanta per second (quanta s^{-1}). The relationship between watts and quanta is not simple, since the energy of a quantum varies inversely with wavelength; quanta from short wavelengths (e.g. blue) have considerably more energy than those from long (e.g. red). Thus the quanta m^{-2} s^{-1} representing a given W m^{-2} will depend on spectral composition. However, Morel and Smith (1974) have shown that for most natural waters there is no more than a $\pm$ 10% variation from the figure given by 2.5 $\times$ 10^{18} quanta s^{-1} W^{-1}, and, for the purposes here, the conventional expression of irradiance in W m^{-2} is sufficient.

The light necessary for photosynthesis arrives at the surface of the sea in amounts which depend upon latitude, season, and cloud cover. It must then pass through the water surface and into the water column. The penetration and transmission of light through the sea also are affected by several factors.

As latitude increases there is greater seasonality of light availability, in terms of watts per square metre per day (W m^{-2} d^{-1}). Figure 2.1 shows that this comes about from winter diminution of light energy at high latitudes – summer light availability is much less variable. Over most of the year daily insolation is lower at higher latitudes, the exception being in summer, when the long days mean that high latitude insolation may exceed levels in the tropics.

The data used to plot Fig. 2.1 ignore losses of light energy due to the atmosphere. These are substantial: Gates (1962) has summarised the fate of the average annual northern hemisphere radiant flux, and the figures are given below.

Reflected by clouds: 25%
Scattered back to space: 9%
Absorbed by atmosphere: 19%

Thus only 47% of the average annual solar flux reaches the earth's surface.

Light reaching the surface consists of both the direct solar beam and light scattered to the surface from the clouds and sky. On average these are in roughly equal proportions, obviously one or the other component will be more

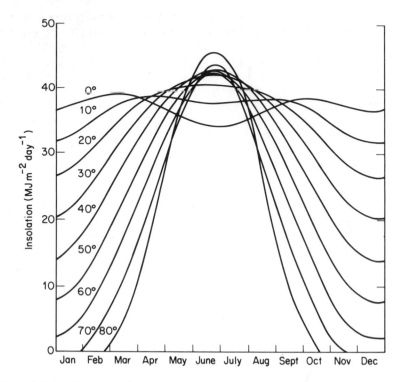

Fig. 2.1   Daily insolation as a function of latitude. (From Kirk, 1983).

important depending upon cloud conditions. A few dispersed clouds may actually increase total irradiance at the surface provided they do not obscure the sun (Kirk, 1983).

The penetration and passage of light

Once it reaches the sea surface, light may be reflected away again, in amounts depending on the angle of incidence. As this decreases, more light is reflected away. Table 2.1 gives calculated values of reflectance for various angles of incidence, showing that reflectance increases markedly at higher levels of $i°$.

In practice the sea surface is never so calm as is assumed for Table 2.1, and the

Table 2.1   Reflectance from a calm surface as a function of angle of incidence. Taken from Jerlov (1968) and Kirk (1983).

Angle of incidence, $i°$	Reflectance (%)
0	2.0
45	2.8
55	4.3
65	8.7
75	21.2
85	58.3
89	89.6

roughening caused by waves complicates matters. Jerlov (1968) stated that waves cause an overall increase of the angle of incidence at high solar elevations (i° small) but that as reflectance is low this effect is unimportant. At low solar elevations (i° large), there is an overall decrease of i° and a corresponding decrease of reflectance thus giving a greater penetration of light. Roughening also decreases reflectance of diffuse light.

'Sea water' is a mix of water, dissolved matter, and particles. Pure water both absorbs and scatters light, and the attenuation of light depends on both processes, increasing as dissolved and particulate materials are added to the mix. Determination of absorption and scattering independently presents problems, for while it is possible to arrange to collect most of the scattered light in the detection system during absorption measurements, experimental measurement of scattering inevitably includes losses due to absorption which must be compensated by applying a correction factor (Kirk, 1983). Scattering intensifies vertical attenuation because scattered photons traverse a greater path length to achieve a given depth, thus they are more likely to be absorbed, and also some will be scattered back in an upwards direction and may pass out of the water column. About half of the total upwelling light flux is reflected downwards again at the air-water interface (Kirk, 1983), the rest emerges and is lost to the sea. The upwelling light beneath the surface may be expressed as the ratio of upwelling to downwelling irradiance, E_u/E_d, called the irradiance reflectance, R. Early estimates of this quantity for oceanic waters lie between 0.5 and 6%, depending on wavelength (Sverdrup, *et al.*, 1942; Jerlov, 1968). R will be greatest in the blue (short wavelength) region of the visible spectrum, as this is where water is most transparent to light. Morel and Prieur (1977) have determined R across the spectrum in a variety of sea water types, ranging from the clear blue of the Sargasso Sea to the more turbid water of upwelling regions. These may be classified into three groups: clear blue water, green water with phytoplankton particles predominating, and green water with inorganic particles predominating. Figure 2.2 shows R across the visible spectrum for

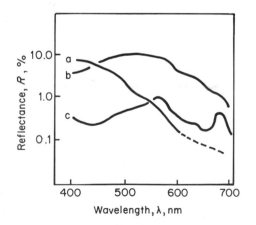

Fig. 2.2 Irradiance reflectance in **(a)** Sargasso Sea water, **(b)** coastal water rich in inorganic particles, **(c)** coastal water rich in chlorophyll. The broken section of line **(a)** is plotted using theoretical rather than observed values. (From Morel and Prieur, 1977.)

examples of these three types. In the blue the values are highest in clear water, intermediate in green water with inorganic particles, and lowest in green water rich in phytoplankton. Nowhere are they higher than ~10%. Therefore, the presence of phytoplankton actually decreases the already small loss of light energy due to scattering, even though the scattering coefficient itself must be higher than in clear water. This happens because of the greater absorption in 'green water' due to chlorophyll.

Pure water absorbs only weakly in the blue and green parts of the spectrum, absorption increases steeply above about 600 nanometres (nm) so that red and infra-red wavelengths are absorbed significantly. Raymont (1980) stated that 98% of total infra-red is absorbed in the top 2 m of the sea. Beyond the other end of the visible spectrum absorption rises again, so that ultra-violet radiation penetrates weakly. In the clearest waters ultra-violet may penetrate to 10 m or more, this is reduced to 1 or 2 m in turbid water (Raymont, 1980).

As the amount of dissolved and particulate matter is greatest near coasts, coastal seas are least transparent to light, while oceanic waters have the greatest transparency. This effect is shown in the Fig. 2.3, where the extinction coefficient (k) per metre at different wavelengths is shown for different water types. The extinction coefficient represents the rate of decrease of radiant energy with depth, it is similar to the absorption coefficient in optics. It may be calculated from:

$$k = \frac{2.3 (\log I_1 - \log I_2)}{d^2 - d^1}$$

where I_1 and I_2 represent irradiance at two depths d_1 and d_2. The factor of 2.3 converts from logarithms base 10 to logarithms base e. The attenuation of light

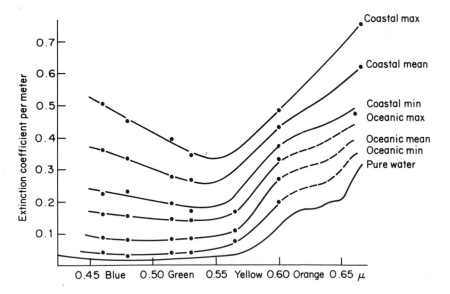

Fig. 2.3 Extinction coefficients of radiation of different wave lengths in pure water and different types of sea water. (From Sverdrup *et al.*, 1942.)

in water is logarithmic, and so it is necessary to use logarithms.

In Fig. 2.3 it is seen that while inshore waters are less transparent across the visible spectrum, the shape of the curve of k on wavelength changes so that they are also relatively more opaque to blue light. This happens partly because of the presence of a group of chemical compounds variously called humic compounds, yellow substance, Gelbstoff, or gilvin. The latter name has been proposed by Kirk (1983), and will be used here.

Gilvin is formed as a result of the decomposition of plant matter, and consists of relatively stable breakdown products of carbohydrates and amino acids. Gilvin does not represent a chemical entity but rather a complex mixture of compounds, which may be present in dissolved or colloidal form. They are known to be formed both terrestrially (in the soil and in freshwater) as well as in the sea, and indeed terrestrial gilvin is of considerable importance in the sea, brought there by land drainage. Sieburth and Jensen (1968) considered that marine and terrestrial gilvin compounds were chemically similar, though much of the gilvin in freshwater precipitated out when in contact with sea water. Using a standard prepared from *Fucus vesiculosus*, they estimated that in bog water there was 17 mg l^{-1} of gilvin, while in river water the concentration was ~ 1 mg l^{-1} and in coastal sea water 0.003–0.8 mg l^{-1}. Marine gilvin was undetectable in the dark winter months, suggesting a close association with phytoplankton activity. Bricaud *et al.* (1981) concluded that material of terrestrial origin was most important in providing optically active gilvin, even in the open ocean. They found that the concentration of gilvin (estimated spectrophotometrically) was unrelated to phytoplankton biomass, and suggested that in oceanic waters gilvin represented a stock of old stable organic material. In inshore waters gilvin concentration was related to freshwater runoff. Gilvin is undoubtedly formed in the sea from plant materials, and may be important locally, especially in regions of high productivity. However it seems likely that, overall, gilvin of terrestrial origin is of greater importance. The close inverse relationship between gilvin concentration and salinity, as has been found for example in the Atlantic west of Ireland (Monahan and Pybus, 1978), is consistent with this interpretation.

The importance of gilvin is in the absorption of light due to its presence. It absorbs particularly strongly at the blue and ultra-violet wavelengths, and therefore reduces the transparency of the sea in that portion of the visible spectrum where water is most transparent (Fig. 2.3). Bricaud *et al.* (1981) considered that even when present in low concentrations in the open oceans, gilvin was as significant in light absorption as a low to moderate algal biomass. Kirk (1983) presented data on absorption, due to gilvin and particulate matter, which show that in Jervis Bay, southeast Australia, gilvin was three times as important as particulate matter. In the Sargasso Sea particles were more important, the value for gilvin being zero. However data of the sort needed for these comparisons are very sparse, and the most that can be said at the present is that gilvin is a significant factor in light absorption in many waters.

Phytoplankton cells themselves absorb light, and where populations are very dense they may limit their own further growth by self-shading. Absorption by phytoplankton will be discussed further below.

The logarithmic attenuation of light vertically in the sea, resulting from the processes discussed above, causes rapid diminution of light energy with depth. Total photosynthetically available radiation is considered to be in the wave-

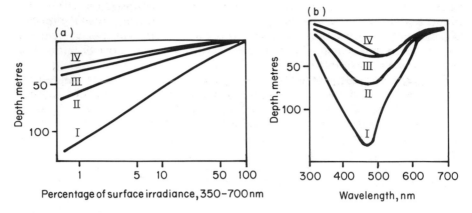

Fig. 2.4 **(a)** Attenuation of visible light with depth in the sea. Water types I, II and III are oceanic, type IV is coastal. **(b)** Depths at which downward irradiance is 1% of the surface value, for sea water classified after Jerlov (1968). (From Jerlov, 1968.)

lengths 400–700 nm (Kirk, 1983). Figure 2.4a shows the percentage of irradiance between 350 and 700 nm at different depths in different water types, from the classification of Jerlov (1968). I, II and III are oceanic water, I being the most clear, IV is clear coastal water. The 1% of surface irradiance value occurs at about 100 m in type I water, diminishing to ~ 20 m in type IV. This depth is known as the compensation depth; as a rough rule, 1% of surface irradiance (assuming full sunshine) is the lower limit for net primary production. Below this level, respiration by phytoplankton cells exceeds photosynthesis and there is no net primary production or accumulation of biomass. Although daylight penetrates to at least 1 km in the clearest ocean water, the euphotic zone is much shallower than this. As would be expected from the remarks above, concerning relative transparency to different wavelengths, the sea acts as a selective filter and the penetration of different wavelengths is very different. Figure 2.4b illustrates this phenomenon, showing that blue light penetrates best in most water types. There is a slight shift in favour of green in more turbid water, because here phytoplankton chlorophyll and dissolved gilvin are absorbing blue wavelengths (see also Fig. 2.3).

While the compensation depth refers to light level as it affects the photosynthesis and respiration of a single cell, the concept of the critical depth is related to total phytoplankton population activity. Recognition of the phenomena involved began with the early studies on phytoplankton production in the Scottish sea loch, Loch Striven. Marshall and Orr (1930) considered that while surface light, nutrients and temperature in winter were all sufficient for diatom growth, because of mixing of the water column cells were frequently carried below their compensation depth. Seasonal overturn of the surface of the water column was discussed in Chapter 1, where it was pointed out that in temperate waters there may be an alternation between a vertically stable, stratified water column in summer and a well mixed one in winter. Marshall and Orr (1930) stated that the regularity of onset of the spring increase of phytoplankton indicated that increasing light must be the triggering factor, but that the extent of vertical mixing was an important modifying influence. Sverdrup (1953) discussed events in a hypothetical water column with phytoplankton evenly distributed with depth. Under such circumstances, production would decrease logarithmically with depth (following the decrease in light availability) but respiration would remain the same. For net production to occur, gross production must exceed respiration.

The graph of gross production and respiration with depth (Fig. 2.5) shows these quantities for the population in terms of area on the graph, the depth at which population photosynthesis equals population respiration is known as the critical depth. This will be substantially deeper than the compensation depth. If vertical mixing is so vigorous that phytoplankters are constantly being lost by being swept down below the critical depth, this loss may exceed any population growth nearer the surface. If the depth of the mixed layer exceeds the critical depth (or exceeds it by some factor) population growth will not take place. Sverdrup was able to test these ideas with data from the Norwegian Sea, from which he could establish the extent of mixing and the critical depth. The results agreed with the idea that the relationship between critical depth and depth of the mixed layer was important in determining overall phytoplankton production.

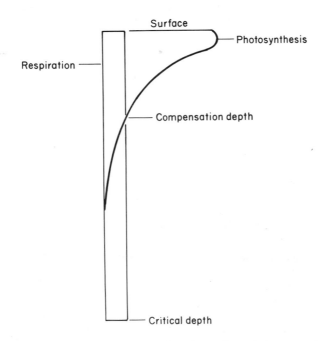

Fig. 2.5 Photosynthesis and respiration in an imaginary phytoplankton population distributed evenly in a water column of uniform temperature with depth. While the rate of photosynthesis varies with depth (following light attentuation), the rate of respiration is uniform. Where photosynthesis per cell is equal to respiration per cell, the individual cell is just in energy balance — this is the compensation depth. Where population photosynthesis is equal to population respiration, as indicated by the areas under the respective graphs, the population as a whole is just in energy balance — this is the critical depth. The critical depth is considered to be some 3 to 6 times the compensation depth. (From Parsons, T.R. and Takahashi, M. (1973). *Biological oceanographic processes*, Pergamon Press, Oxford.)

Physiological aspects

A photon of wavelength 700 nm (red) contains only 57% as much energy as one of 400 nm (blue). This may be estimated from the expression $\epsilon = (1988/\lambda) \times 10^{-19}$ J, where λ is the wavelength in nm (Kirk, 1983). It might therefore be concluded that blue light is more valuable for photosynthesis than red, however this is not so, for photosynthetic yield depends on the number of quanta captured rather than on the energy per quantum. Red light is therefore more efficiently converted than is blue light, since the higher energy of blue quanta is not used.

As blue light penetrates best in water, pigments absorbing in these wavelengths may perhaps be especially valuable for algae at depth. McCarthy and Carpenter (1979) suggested that the cyanobacterium (blue-green alga) *Oscillatoria thiebautii* may be at a disadvantage at depth, because of poor absorption of blue light. Shimura and Fujita (1975) argued to the contrary, pointing out that phycoerythrin in this alga has a triple absorption peak at 500, 547 and 565 nm, which they consider gives it at least adequate absorption capacity in the blue wavelengths.

More generally, Yentsch (1980) drew attention to the distribution of peaks of absorption by major algal pigments in the 'clear windows' of the spectrum of light transmitted through water (Fig. 2.6). The chlorophylls have major absorp-

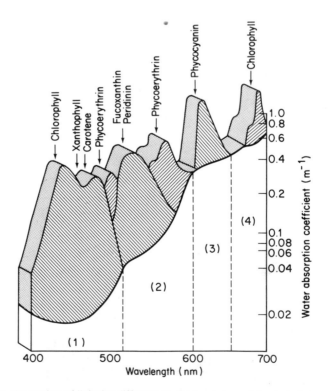

Fig. 2.6 The absorption of light by different algal pigments and the 'windows of clarity' in water absorption. The spectra for the pigments approximate to those measured *in vivo*. (From Yentsch, 1980.)

tion peaks in the 425–450 nm band, with another in the 665–680 nm band. In between these are the absorption peaks of the carotenoids of diatoms and dino-flagellates, and the phycoerythrins and phycocyanins of cyanobacteria. Although chlorophyll *a* is the only pigment actually to carry out the primary photochemical reactions of photosynthesis, light energy may be absorbed by other pigments and passed to chlorophyll *a* (Brody and Brody, 1962).

The pigments involved in photosynthesis are divided between two photo-systems, called I and II. Both photosystems have chlorophyll *a* associated with them, photosystem II (PS II) is responsible for photolysis of water and photo-system I (PS I) for the production of the high energy compounds required for the reduction of carbon dioxide (Jørgensen, 1977). In diatoms, the carotenoid fucoxanthin and chlorophyll *c* are associated with PS II, β-carotene with PS I. Much of the chlorophyll *a* in diatoms is associated with PS II (unlike the situa-tion in most plants, where PS II has about one third of the chlorophyll *a* and all the chlorophyll *b*). However this is not constant even within one species: the chlorophyll in PS I increases with age of the cell, and also experimentally under low light intensities or in red light (Jørgensen, 1977). The two systems are to be seen as operating in series, both dependent on energy coming ultimately from light (Butler, 1978). If PS II is 'overdriven' excess energy can be diverted to I, and *vice versa*. Chlorophylls are non-covalently bound with proteins in both PS I and PS II, and the nature of the complex in PS I merits brief discussion here. An entity has been identified which is called the chlorophyll-P700 complex (Thornber, 1975), in which P700 is the reaction centre chlorophyll for PS I. It consists of a chlorophyll *a* dimer. As the reaction centre, it occupies a central position in the whole photosynthetic process. The light-harvesting pigments may be seen as an antenna which collects light energy, passing it on to the reac-tion centre of P700. In so-called shade adaptation, the size of the antenna may increase; this can be measured as a change in the ratio of antenna pigments to P700. This will be mentioned again below and in Chapter 3.

There is an optimal light intensity for photosynthesis, and at both higher and lower intensities photosynthesis falls from the maximum. These features are

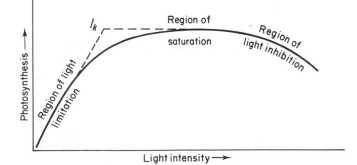

Fig 2.7 General curve relating photosynthesis to light intensity. The curve shows that photosynthesis initially rises with increasing light, reaches saturation, and then at very high intensities begins to decline. The point I_k, which may be estimated by extrapolation from the 'limited' and 'saturated' portions of the curve, was proposed as an 'index of saturation'. (From Talling, J.F. (1957), *Photosynthetic characteristics of some freshwater plankton diatoms in relation to underwater radiation*, New Phytologist, **56**, 29–50.)

shown in Fig. 2.7. The so-called photoinhibition at high light intensities is responsible for the decrease in photosynthesis often reported for the extreme surface layer, however experiments in which algae may be subjected to high light intensities for longer than they would be in the water column will overestimate the importance of the phenomenon. Photoinhibition occurs because of damage to pigments at high light intensities. In the freshwater diatom *Asterionella formosa* it has the characteristics of photooxidative damage to both light and dark reaction mechanisms (Belay and Fogg, 1978), with shorter wavelength radiation being the most damaging. This is probably representative of phytoplankton in general. In *A. formosa* inhibition was small during the first hour, becoming significant in the second hour, in a lake experiment (Belay, 1981). These time scales mean that in the normal somewhat turbulent water column, cells may not spend long enough near the surface to be significantly affected. Freshwater phytoplankton populations show evidence of adaptation to resist photoinhibition when the water column is stable (Belay, 1981).

There is some adaptation by cells to light intensity, as has been indicated above in the discussion of photosystems. In diatoms, such adaptation may be morphological or physiological (Jørgensen, 1977): *Nitzschia closterium* is

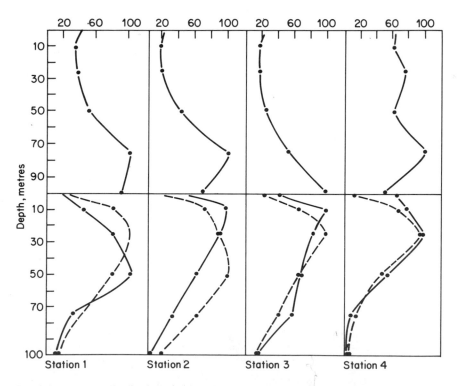

Fig. 2.8 Vertical distribution of chlorophyll (upper graphs) and carbon fixation (lower graphs) at four stations in the Sargasso Sea. For carbon fixation, both *in situ* photosynthesis measurements (solid lines) and indirect estimates based on chlorophyll levels and light availability (broken lines) are given. The common x axis is a percentage scale. (From Yentsch, 1974.)

reported to increase cell volume with decreasing light intensity, while chloroplast size decreases with increasing intensity. In terms of overall production levels, dark or shade adaptation in marine phytoplankton is said to be unimportant (Yentsch, 1974), for in spite of the common occurrence of a deep chlorophyll maximum above the base of the euphotic zone there is so little light here that photosynthesis is much reduced (Fig. 2.8).

In the laboratory, algal cells may show a lag of about 24 hours in adaptation on transfer from low to high light intensities (Jørgensen, 1977). Falkowski (1983) used the chlorophyll/P700 ratio to investigate the illumination history of natural phytoplankton. From data on the rate at which the ratio alters during shade adaptation and the variations in the ratio through the euphotic, it was shown that the scale of adaptation time *vs* mixing time varied depending on location and season. Lewis *et al.* (1984) have concluded that turbulent mixing is an important factor in controlling phytoplankton productivity. If the time scale of mixing is shorter than that of adaptation, phytoplankters will not be able to adapt effectively to any given light level. This will be referred to again in Chapter 3.

Nutrients

Phytoplankton algae require both inorganic and organic substances for growth, the former have been more extensively studied. Some of these substances are available to excess, many others are not, notably those ions supplying nitrogen, phosphorus and silicon.

The principal nutrient elements

Nitrogen is present in sea water in a wide range of forms, from dissolved molecular nitrogen to inorganic and organic compounds. Fixation of molecular nitrogen by cyanobacteria may be a significant process in the open ocean (Carpenter and Price, 1977). The most important combined form is the nitrate ion (NO_3). Ammonia (NH_3) is an important excretory product of zooplankton, it is readily oxidised to nitrite (NO_2) which in turn is oxidised to nitrate. In completely anoxic conditions, reduction of nitrate to ammonia or nitrogen may occur through bacterial denitrification. All three forms may be used by phytoplankters.

Dissolved phosphorus occurs mainly as inorganic orthophosphate (PO_4^{3-}), usually simply 'phosphate'. The chemistry of phosphorus in the sea is simpler than that of nitrogen, and because of its relative ease of analysis it was the first nutrient to receive close attention.

The cycle of silicon involves exchanges between silicic acid in solution ($Si(OH)_4$) which is mainly un-ionised (Spencer, 1975), and various mineral compounds. Diatoms and radiolaria require silicon for mineralised cell walls, deposited as silica (SiO_2). Darley and Volcani (1969) state that silicon is practically ubiquitous in biological material, but that an absolute requirement for it is not. They showed that in the diatom *Cylindrotheca fusiformis* silicon is necessary for synthesis of deoxyribonucleic acid, quite apart from its role in the cell wall.

The distribution of nutrients

It has been recognised for some time that the principal nutrients show a characteristic depth distribution in the oceans, being depleted in surface waters and abundant at greater depths (Fig. 2.9). The reason for this distribution is the uptake and incorporation of nutrients by phytoplankton in the surface illuminated layers, and their subsequent removal to deeper water by death and sinking of phytoplankton or by zooplankton which have been feeding near the surface. There is some recycling in the surface layers, and this is an aspect which has received increasing attention, but overall the major phenomenon is one of net surface depletion. The small fluctuations at depth indicated in Fig. 2.9 may be related to deep water circulation, as are the differences evident between the oceans.

Nutrients do not accumulate indefinitely in the deep ocean because of an ocean-wide very slow upwelling of deep water. Bottom water is formed only at very high latitudes (see Fig. 1.4), it spreads out through the oceans at depth and very slowly returns towards the surface. It has been calculated that on an ocean-wide basis, this involves the upward movement of the equivalent of a layer 2 m thick in one year. Therefore it is much slower than the processes of coastal or equatorial current upwelling discussed earlier (Chapter 1), sometimes called enhanced upwelling. There are also spatial differences in nutrient concentration, an example is shown in Figs. 1.8 and 1.10 where upwelling water (enhanced upwelling) is seen in the distribution of phosphate and silicate.

The features indicated in Fig. 2.9 are very general, and typical of the open ocean. Near coasts surface nutrient levels are generally elevated due to the effects of land drainage, evidence of this can also be seen in Fig. 1.8. However even here there is a tendency for surface depletion, particularly during periods of phytoplankton activity. Data for the English Channel station E1 over 16

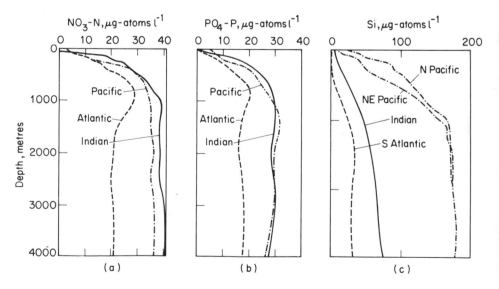

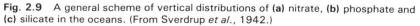

Fig. 2.9   A general scheme of vertical distributions of **(a)** nitrate, **(b)** phosphate and **(c)** silicate in the oceans. (From Sverdrup *et al.*, 1942.)

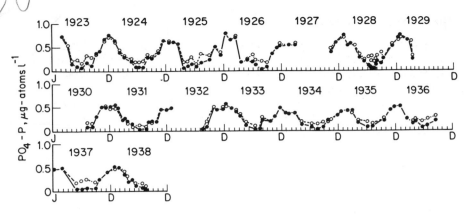

Fig. 2.10 Concentrations of phosphate-phosphorus at the surface (solid line) and near the bottom (broken line) at station E1. (From Cushing, 1975.)

years show a regular cycle in phosphate levels at both the surface and near-bottom (70 m) depths (Fig. 2.10). The data for E1 indicate depletion throughout the water column, and we now know that this location is less well stratified than further west, due to tidal mixing (Fig. 1.12). The clear coincidence of phytoplankton activity with phosphate depletion was suggestive not only of a link between them, but also perhaps that phosphate depletion might limit production levels after the initial spring bloom. Cooper (1938, 1948) in particular argued that there was a link between winter phosphate levels and the subsequent level of phytoplankton production, and that this was further to be seen in the eventual landings of fish. Atkins (1925) showed that the surface water above the seasonal thermocline became particularly deficient in phosphorus in summer, and suggested that stratification, by inhibiting mixing, prevented use of nutrients from deeper water.

In more recent years attention has turned to nitrogen as a potentially limiting nutrient element, and some authors consider that in the sea nitrogen is almost always limiting. Goldman (1976) pointed out that in coastal phytoplankton species the ratio of nitrogen to phosphorus by atoms was between 10:1 and 20:1, while in the water it was usually 5:1. Enrichment with nitrogen enhanced algal growth, which was consistent with the supposition that it was a limiting factor. McCarthy (1980) was more cautious about the limiting role of nitrogen in the sea, pointing out that natural assemblages of phytoplankton often grow at rates permitting doublings every 2–4 days while 'nutrient concentrations border on the limits of analytical detection'. This will be discussed further below.

Silicon is usually available in apparent excess (Paasche, 1980). However, particularly in regions of upwelling it may become depleted before nitrogen or phosphorus, and this may limit diatom blooms in upwelled water. It has been suggested that the relatively weakly silicified summer forms of many common temperate phytoplanktonic diatoms may be a response to silicon shortage.

As well as the so-called micronutrients dealt with above, trace metals (e.g. iron, copper, manganese) are known to be necessary for phytoplankton growth. Some species require the presence of one or more of the B vitamins cyanocobalamin, thiamin or biotin (Swift, 1980). These are present in sea water in very low concentrations, released by both phyto- and zooplankton and by bacteria.

Physiological aspects

Nutrient distribution is clearly much more under the influence of planktonic organisms themselves than is the distribution of light. Indeed, in the case of dissolved vitamins, the substances are entirely the result of life processes.

Uptake kinetics

Dugdale (1967) discussed the role of nutrients as limiting agents in phytoplankton growth. He made use of the Michaelis Menten expression describing the rate of enzyme action, namely:

$$v = \frac{V_m S}{K_s + S}$$

where
v = rate of action
V_m = maximum rate
S = substrate concentration
K_s = half-saturation constant

There is evidence that this expression can describe both the uptake of a nutrient and the growth of algae in relation to nutrient concentration (Dugdale, 1967; Eppley and Thomas, 1969; MacIsaac and Dugdale, 1969). Depending on the values of V_m and K_s, different algae might be at an advantage in high or low nutrient concentrations. Dugdale (1967) showed this for two diatom species using known values for V_m and 'hypothetical' values for K_s. If species with low V_m values have low K_s values, they will be able to compete effectively at low concentrations of nutrients. MacIsaac and Dugdale (1969) found evidence that there was such an association between V_m and K_s in natural oceanic phytoplankton, so that different species would be favoured depending upon whether nutrients were impoverished or not. Eppley *et al.* (1969) showed that small-celled oceanic species in general had the lowest K_s values (see also Malone, 1980). They extended the study to consider light and temperature, pointing out that K_s values by themselves have little meaning in ecological terms. By taking these other variables into account as well, they were able to offer some explanations for the distributions of four algae: *Coccolithus huxleyi* (oceanic), *Skeletonema costatum* and *Ditylum brightwelli* (coastal), and *Dunaliella tertiolecta* (rock pool). Figure 2.11 shows that *C. huxleyi* should compete best at all levels of nitrate under relatively low light conditions, with higher irradiance it is still superior at low nitrate levels while *S. costatum* and *D. brightwellii* become superior as nitrate increases. *D. tertiolecta* always did poorly in comparison. Although Fig. 2.11 shows growth in response to nutrient availability the response of v in the Michaelis Menten equation is known to be the same, so that the work of Eppley *et al.* (1969) illustrates the phenomenon to which Dugdale (1967) alluded. Other authors (e.g. Friebele *et al.*, 1978) considered that the relationship between uptake kinetics and cell size is unclear and confusing, and that generalisations are not possible. This will be discussed further in Chapter 3.

McCarthy and Goldman (1979) have shown that the diatom *Thalassiosira pseudonana* is able to elevate V_m in response to nitrogen shortage, and suggest

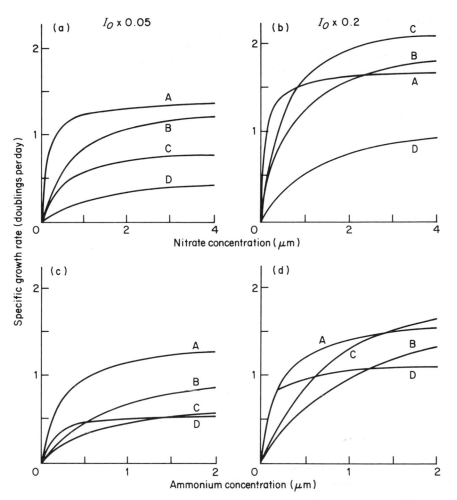

Fig. 2.11 Specific growth rates of four planktonic algae in relation to nitrogen availability five per cent at **(a)** 5%, **(c)** 5%, **(b)** 20%, and **(d)** 20% of surface light intensity (I_0). **(a)** and **(b)** nitrogen as nitrate; **(c)** and **(d)** as ammonium. The algae are A, *Coccolithus huxleyi*; B, *Ditylum brightwelli*; C, *Skeletonema costatum*; and D, *Dunaliella tertiolecta*. (From Eppley *et al.*, 1969.)

that this offers a resolution of the paradox where phytoplankton are able to grow at very low apparent concentrations of nutrients. Phytoplankters are seen to be somewhat opportunistic in their ability to take up nutrients, and so make maximal use of tiny and short-lived 'patches' of nutrients encountered in an environment which is on the whole impoverished. These patches would arise from the activity of zooplankters or bacteria (see below), and would be unpredictable in time and space. There is evidence that such a micro-scale patch structure of nutrient distribution does exist (McCarthy, 1980).

Observations such as these make it more difficult to believe that there might be a simple, clear-cut relationship between nutrient level and primary productivity. At the same time, the suggestion that smaller cells may be better able to make use of nutrients in low concentrations offers at least a partial explanation of the decrease in cell size along the gradient of diminishing nutrients from coastal waters to the open ocean (Yentsch, 1974). When nutrients are limiting, it is most likely to be nitrogen which is involved as the limiting element.

Nutrient regeneration

Although there is a tendency for a downward flux of nutrients out of the surface layers, there is considerable cycling of nutrients within the euphotic zone. As phytoplankton cells are eaten, they may be ruptured, releasing cell contents. More importantly, zooplankters replenish levels of phosphorus and nitrogen through excretion. This may be of importance in maintaining primary production after the initial outburst of the spring bloom. For example, Holligan *et al.* (1984b) estimated that > 50% of the nitrogen requirement of phytoplankton in stratified waters in the English Channel in summer might be supplied in this way. This is about the same fraction as is said to be supplied by regeneration of nutrients in upwelling waters (Harrison, 1980). Thus even in these productive waters, regeneration of nutrients in the euphotic zone may be important in sustaining phytoplankton growth.

A combination of ideas on patchiness and nutrient cycling may explain the phenomenon of continued phytoplankton growth in oligotrophic waters where nutrients are scarce (Goldman, 1984). It is suggested that the tiny aggregates of amorphous organic matter ('marine snow') commonly found in the water column are sites of high levels of microbial activity and primary production, with autotrophs and heterotrophs in close proximity.

Nitrogen fixation

One of the interesting aspects of the nitrogen cycle in the sea is the fixation of elemental nitrogen. This is brought about by prokaryotic organisms, including not only bacteria but also those formerly known as the cyanophyceae or blue-green algae, now called cyanobacteria. The enzyme nitrogenase catalyses the reduction of nitrogen to ammonia, it is deactivated by oxygen and many cyanobacteria have special cells called heterocysts which seem to offer protection to the enzyme (Bothe, 1982). The planktonic cyanobacterium *Trichodesmium* lacks heterocysts, but is known to fix nitrogen. The organism occurs in colonies made up of several hundred strands, each strand containing about 100 cells (Fogg, 1982). Strands passing through the colony centre have been found to contain lightly pigmented cells in a central region, these do not photosynthesise and seem to offer an environment for the action of nitrogenase (Carpenter and Price, 1976). Experimental disruption of the colonies sharply diminished nitrogen fixation; if disruption by wave action has the same effect this could be an explanation for the observations that *Trichodesmium* thrives under calm conditions. While nitrogen fixation is probably very important for the organisms responsible, in terms of the overall nitrogen budget it is of unknown significance. In the Caribbean Sea, fixation has been estimated to be 1.3 mg N m^{-2} d^{-1} (Carpenter and Price, 1977). This is an average value for a number of stations in the eastern portion of the Caribbean, and it represents 8% of the total phytoplankton demand for nitrogen in this region. However the loss of nitrogen from the euphotic through sedimentation has been estimated as 6.22 mg N m^{-2} d^{-1}, and the fixation is about 20% of this value; on this basis it represents a significant input. This is exceptional: other studies suggest much lower levels of nitrogen fixation, as in the Sargasso Sea (Carpenter and Price, 1977) and the central north Pacific (Mague, *et al.*, 1974). The ecological significance of this remains uncertain: supply of 'new' nitrogen by fixation in the stratified surface layers of the oceans may be of greater importance than is indicated by the daily fixation values.

3

The autotrophic plankton

The word 'plankton' comes from the Greek planktos, a wanderer. This refers to the fact that planktonic organisms are drifters rather than powerful swimmers, their horizontal distribution is more governed by currents than by the outcome of their own efforts. Nevertheless as we shall see, vertical zonation in the sea is a conspicuous feature of planktonic distributions. Some of this has already been alluded to in the occurrence of a chlorophyll maximum in the thermocline (Chapter 2).

The rather cumbersome term 'autotrophic plankton' has been used to head this chapter in preference to the more widely accepted 'phytoplankton'. This is in order to draw attention to the ecological role of the organisms under consideration, because the classical taxonomic divisions are secondary here. Many of the phytoplankton are largely or wholly heterotrophic, and mention of some of them will be made later (Chapter 5). Moreover, the Cyanobacteria (= Cyanophyceae, blue-green algae) are now accepted as belonging to the prokaryotes, far more closely allied to the bacteria than to the algae. Yet they are at least mainly autotrophic, while the bacteria are mainly heterotrophic. The taxonomic difficulty is avoided by using a functional classification. Nevertheless, the term phytoplankton is a useful one, and it will be used where appropriate.

There are a great many autotrophs in the plankton, and their taxonomy is complicated. The most prominent are the Bacillariophyceae or diatoms, the Dinophyceae or dinoflagellates, and the Cyanobacteria. In size these range from less than 1 μm to about 2 mm, and they may occur as single cells, or as colonies. In addition to these forms, there is a host of often smaller algae, many motile by means of flagella, placed in the Chlorophyceae (e.g. *Dunaliella*), Haptophyceae (e.g. *Coccolithus*), Chrysophyceae, Cryptophyceae, Xanthophyceae and Prasinophyceae. Although these have been the object of recent taxonomic studies less is known of their ecology than, for example, that of the diatoms and dinoflagellates. They are often referred to collectively simply as 'flagellates'.

The taxonomic divisions used above are those of Parke and Dixon (1976); other authorities will differ.

The organisms

Bacillariophyceae DIATOMS

The diatoms (Fig. 3.1) are ubiquitous and prominent autotrophs. A major characteristic is the cell wall, consisting of both silica and organic components. The cells are variable in shape between and also within species, and may also vary in

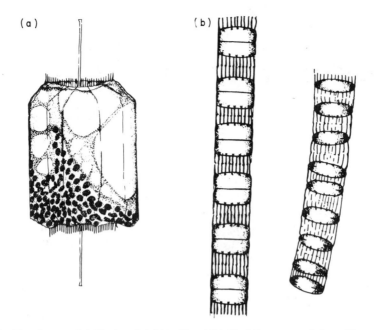

Fig. 3.1 The diatoms **(a)** *Ditylum brightwelli* and **(b)** *Skeletonema costatum*. (From Hendey, N.I. (1964). *An introductory account of the smaller algae of British coastal waters. Part V: Bacillariophyceae (diatoms)*, Fishery Investigations Series IV, HMSO, London.)

size within species. Their colour is predominantly green or brown, and nutrition is almost always autotrophic by photosynthesis. Motility is very limited; some species show a gliding movement when adjacent to a solid object, though the agency of motion is obscure. This behaviour can be seen among planktonic diatoms only in certain sorts of colonies when the cells exist side by side and may move relative to one another. Any control the cells exercise over position in the water column must be mediated by regulation of buoyancy (see below).

Dinophyceae

The dinoflagellates (Fig. 3.2) in contrast are motile, possessing two flagella. One is longitudinal and is a conventional flagellum, the other is transverse and has a ribbon-like form. This transverse flagellum usually lies in a transverse circumferential furrow, the girdle. There is an outer cell wall membrane, in the so-called armoured genera thecal plates lie under this. About half the dinoflagellates have chloroplasts and are autotrophic, the remainder lack chloroplasts and are heterotrophic. An account of their biology is given by Spector (1984).

Dinoflagellates have considerable powers of locomotion, and they may use this ability to choose certain levels in the water column. An outcome of this behaviour may be the formation of a so-called red tide. These are features of coastal waters and are more frequent in the tropics and subtropics than temperate seas, though red tides are known to occur in the North Sea. In favourable conditions, which must include calm weather, dinoflagellate cells may

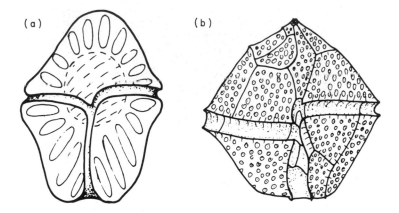

(a) (b)

Fig. 3.2 The dinoflagellates **(a)** *Gymnodinium splendens* and **(b)** *Gonyaulax polyedra*.
(From Dodge, J.D. 1982. *Marine dinoflagellates of the British Isles*, HMSO, London.)

congregate in a relatively thin layer near the surface. The individual cells are usually yellow-brown, but when massed together in this way they change the water colour to red-brown. The species involved may often include one or more of the toxic forms. *Gymnodinium breve* for example produces a substance which is toxic to fish and in episodes of red tide sufficient of this toxin may be secreted into the water to cause fish kills. Species of *Gonyaulax* on the other hand have toxins which are evident only after they have been passed up the food chain, from filter feeding herbivores to first carnivores, and ultimately to predators, including man.

Cyanobacteria *prokaryotes; fix N.*

Traditionally these have been regarded as algae. It is now accepted that a fundamental division of living organisms can be made at the cellular level, namely the classification into prokaryotes and eukaryotes. Prokaryotic cells are characterised by a simpler level of organisation than is found in eukaryotic cells. Particularly, they lack a nucleus and nucleolus, and do not have mitochondria or chloroplasts. The Cyanobacteria are clearly prokaryotic, while the true algae are eukaryotic. The 'blue-greens' have accordingly been recognised as constituting a group of organisms which are generally photosynthetic, but to be separated from the algae. As has already been remarked, they are able to fix elemental nitrogen.

Fogg (1982) commented that despite the sporadic abundance of blue-greens in the plankton, they are generally not prominent. The most conspicuous is *Trichodesmium*, recognised by Fogg (1982) among others as a separate genus, but often made synonymous with *Oscillatoria*. It may be separated only on the basis of the habit of colony formation. Van Baalen and Brown (1969) described the gas vacuoles found in the cells of *T. erythraeum*. These are present as tiny cavities in the cytoplasm, about 85 nm wide and ten times as long, arranged in regular arrays of two concentric layers of vacuoles. Their walls are of protein (Walsby, 1978). Van Baalen and Brown (1969) suggested that they may have a function in shielding the photosynthetic apparatus of the cell, most of which lies

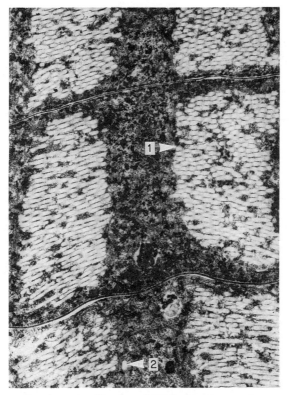

Fig. 3.3 Median section through a filament of *Trichodesmium erythraeum*. The micrograph shows the concentric arrangement of gas vacuoles (arrow 1) lying outside the central portion of the cell with its photosynthetic lamellae (arrow 2). (From Van Baalen and Brown, 1969.)

in the central core inside the layers of gas vacuoles. Figure 3.3 shows these arrangements.

The evidence for nitrogen fixation by *Trichodesmium* has already been discussed. Nitrogen fixation is also carried out by *Richelia intracellularis*, usually found as an endosymbiont of diatoms such as *Rhizosolenia* species, although it may be free living. Unlike *Trichodesmium*, *Richelia* possesses heterocysts. The overall significance of nitrogen fixation in the oceanic nitrogen cycle remains uncertain (see Chapter 2), but there is evidence that fixation by *Richelia* may be locally significant. Venrick (1974) suggested that it may be important in supplying nitrogen and promoting blooms of phytoplankton in the oceanic Pacific.

The significance of the cyanobacteria in planktonic primary production has been thought to be low (Fogg, 1982). However, a problem arises which applies to all such judgements: in the case of very small organisms, their contribution may be seriously underestimated. This happens because small cells may escape through the filters used to retain cells after incubation, in ^{14}C experiments. Johnson and Sieburth (1979) have pointed out that the so-called chroococcoid cyanobacteria, including very small cells of the order of 1 μm diameter, are common and widespread. In the Atlantic, significant populations extend from the waters offshore Iceland to the tropics. Their contribution to the photosynthetic biomass was not estimated, but it was suggested that it might provide at least a partial answer to the problem of underestimation of carbon flow which occurs with the ^{14}C technique. Many authors, for example Platt *et al.* (1983), now consider that the very small autotrophs may make a significant contribution to total carbon fixation.

Buoyancy

For photosynthesis and growth, photoautotrophs need to be maintained in the euphotic zone. Though they do not need to spend all their time there, a cell permanently out of the euphotic will not survive indefinitely unless it is capable of heterotrophy. The problem of suspension is therefore a real one, moreover as there will be a depth above or below which light or some other physical variable is less than optimum the control of buoyancy is desirable. How may flotation be achieved?

The specific gravity of sea water ranges between 1.021 and 1.028, that of cell cytoplasm between 1.03 and 1.10 (Smayda, 1970). Cell wall materials have higher values still, ranging from 1.5 (cellulose) through 2.6 (silica) to 2.95 (calcium carbonate). Clearly cells will tend to sink. In fact the observation is that in a sample of freshly collected phytoplankton, some cells do sink, some remain suspended at one depth, and yet others float. Different cells of a single species may do any one of these, depending upon the physiological state of the cell. It is therefore suggested that cells may be able to regulate their density, and this does seem to happen. In the case of *Trichodesmium* the robustness of the gas vacuoles probably does not allow them to be used in fine regulation of density – they simply provide buoyancy – and it is not known how density may be regulated. The alga is distributed through the water column from the surface to at least 200 m, with a likely optimum depth of about 15 m (Walsby, 1978). The robust vacuoles resist collapse, withstanding pressure well, to at least 120 m, and providing buoyancy for a return to shallower waters. Among diatoms there is an overall inverse relationship between size and sinking rate at least among some forms, but this is very plastic. For example comparing *Thalassiosira nana* and *Rhizosolenia setigera*, Smayda (1970) pointed out that with cell volumes differing by 3200 times, the two have similar sinking rates when in active growth. Moreover, in the species *Skeletonema costatum*, *Rhizosolenia setigera* and *Bacteriastrum hyalinum*, the same cells may show 2 to 5 fold changes in sinking rates in 24 hours. These observations, together with the phenomenon of faster sinking in senescent or dead cells compared with healthy, active ones, suggest the existence of physiological controls of sinking rates. Among diatoms, Anderson and Sweeney (1978) have shown that *Ditylum brightwelli* changes its ionic composition during a period of illumination, becoming more dense, so that the sinking rate gradually increases. The alga was cultured in a regime of 8 h light:16 h dark. During the light phase the Na^+ and K^+ balance changed, favouring K^+. As K^+ is the heavier ion, the sinking rate increased during illumination. Towards the end of the 8 h illumination, sinking rate decreased again and was low in the dark period. In the dinoflagellate *Noctiluca scintillans* (= *N. miliaris*) there is good evidence for regulation of the ionic composition of the cytoplasm, by excluding heavy divalent ions (especially

SO_4^{2-} and accumulating Na^+ in preference to K^+, as well as NH_4^+ (Smayda, 1970). It must be noted that *Noctiluca* is not autotrophic, and feeds by ingesting diatoms. *Pyrocystis noctiluca* is another dinoflagellate, in this case autotrophic but non-motile, and it achieves positive buoyancy by manipulation of ionic ratios (Kahn and Swift, 1978).

Other aids to flotation have been postulated, such as regulation of size or shape. The larger the body, the relatively smaller the surface area with respect to volume, and thus large bodies have relatively lower drag. Small diatoms will sink more slowly than large ones in laboratory experiments (Smayda, 1970). The same phenomenon is said to confer advantages from colony formation in *Trichodesmium*: the near-spherical colonies will rise through the water column faster than would chains of cells (Walsby, 1978). Conversely, in diatoms colony formation increases sinking rate, except in *Skeletonema* where special turbulence conditions may be set up by the numerous silica rods connecting the cells of the chains (Smayda, 1970). It was supposed that this turbulence increased the frictional drag of the sinking chains, and decreased their sinking rate.

The gross shape of the cell undoubtedly affects the sinking rate, especially at smaller sizes. However there is no good relationship between shape, size and flotation of viable cells (Smayda, 1970). Although protuberances such as the siliceous setae of many diatoms have been cited as aids to flotation, the increased density caused by the added silica may outweigh any advantage. It can be shown that spineless forms of the diatom *Rhizosolenia setigera* usually sink faster than spinous forms, this seemed to be a consequence of the spines guiding sinking cells into such a posture that maximum drag resistance is obtained (Smayda, 1970).

Although the density of phytoplankton cells is usually greater than that of sea water, and sinking is therefore in a sense inevitable, it may be wrong to assume from this that flotation requires special morphological or physiological adaptations. The reason for this lies in mixing in the euphotic zone, principally due to Langmuir circulations (Chapter 1). The important point here is that not only are there regular downwellings of water, but also regular upwellings, and if the wind is strong these circulations may extend through much of the mixed layer, or perhaps all of it. Maximum sinking rates for living, active diatoms in the laboratory range from 2.6 to 26 m d^{-1}, with the exception of the very large species *Ethmodiscus rex* (Smayda, 1970). For phytoplankton in the sea it may be better to adopt an 'average' sinking rate of no more than 1 m d^{-1} (Smayda, 1970; Bienfang, 1980; Sournia, 1982). Downwellings in Langmuir convections range between 2.6×10^3 to 5.2×10^3 m d$^-$, upwellings are presumably at least of the same order. These rates of flow are much greater than the 'average' sinking rate, and suggest that active Langmuir circulations will retain particles very successfully. Rapidly sinking particles will fall through the helical vortices, merely being deflected by the water flow. However as sinking rate decreases with respect to circulation rate, the entrainment of particles in the circulation cell will increase. This is represented in Fig. 3.4, where the value for R represents the ratio, sinking speed:circulation speed. It may be concluded that in prolonged calm conditions cells may indeed sink through the euphotic zone and be lost, but that as turbulence increases Langmuir circulations will act to retain a good proportion of phytoplankters near the surface.

The question of the retention, or otherwise, of phytoplankton populations in

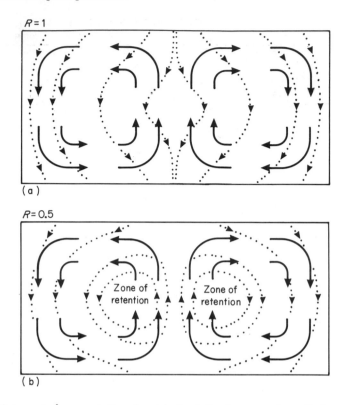

Fig. 3.4 Schematic representation of water circulation in convection cells (heavy lines and large arrows) with the motion of small particles sinking through the cells (dotted lines and small arrows). When the sinking speed and maximum upwelling speed are equal **(a)**, particles sink through the cells, merely being deflected. In this drawing the ratio of the two speeds (*R*) is 1. When the sinking speed is less than the maximum upwelling speed there is a zone of retention which will trap particles entering it **(b)**, for *R* = 0.5, represented by closed particle trajectory lines. The size of the zone will be inversely related to *R*. (Redrawn with modifications from Stommel, H. (1949). *Trajectories of small bodies sinking slowly through convection cells*, Journal of Marine Research, **8**, 24–29.)

the euphotic zone has been investigated at sea. Falkowski (1983) has shown that, depending on the strength of mixing, phytoplankters may or may not be circulated through the mixed layer. On the Georges Bank off Newfoundland, strong tidal mixing throughout the year maintains a turnover of water, nutrients and phytoplankton cells in the euphotic. In contrast, in the waters of the New York Bight mixing is less constant and less predictable, and there is more chance of phytoplankters sedimenting out of the euphotic. Evidence that this does occur in the Bight has been given by Malone *et al.* (1983). Bienfang (1980) determined sinking rates of phytoplankters from the subtropical Pacific (Hawaii). Natural populations from three depths (24, 40 and 71 m) were filtered to obtain two size categories, 3–20 μm and 20–102 μm (these are the size categories as defined by Bienfang). It was found that sinking rates were higher in the larger size fraction, and lower in the smaller one (Fig. 3.5). Sinking rates were lowest for both size categories at 71 m, just above the chlorophyll maximum observed

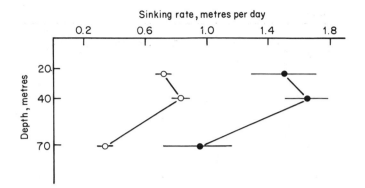

Fig. 3.5 Sinking rates of phytoplankton in the cell size categories of 3–20 μm (open circles) and 20–102 μm (solid circles) at three different depths; the graph shows the mean and standard deviation of triplicate trials for each depth. Sinking rates at the upper two depths were statistically similar within both size categories, while the rate at 71 m was significantly less (P<0.01, Students' t). (From Bienfang, 1980.)

in this study. This is consistent with the hypothesis that the chlorophyll maximum widely observed in the sea and lying somewhere near the base of the euphotic zone is due to increased buoyancy of cells at these depths. At station E1 in the English Channel, there is a seasonal change in the depth distribution of chlorophyll, associated with seasonal changes in stratification and stability in the water column (Fig. 3.6). Over summer when there is a seasonal thermocline, chlorophyll concentration is greatest in the region of the thermocline (Pingree *et al.*, 1977; Holligan and Harbour, 1977).

The sinking rates from laboratory measurements given by Smayda (1970) of 2.6 to 26 m d^{-1} are greater (to an order of magnitude) than those reported by Bienfang (1980) for natural phytoplankton. Greater still are the speeds of current flow in Langmuir downwellings (see above), while the more general vertical movements inferred by Falkowski (1983) range between 0.9 and 86 m d^{-1}. The effects of mixing processes, together with the ability to regulate buoyancy, explain the distributions of phytoplankters described above.

If buoyancy can be regulated, why are the organisms not neutrally buoyant all the time? In part the answer may be that for non-motile cells, slow sinking is an advantage as it prevents local impoverishment of nutrients. If a cell takes up nutrients faster than they can be supplied by diffusion, a local shell of impoverishment will build up round the cell. In Chapter 2, the micro-distribution of nutrients was referred to in terms of the distribution of small patches of enrichment (McCarthy, 1980). Here, the opposite effect is under consideration. Smayda (1970) has argued that sinking is used by algal cells to gain access to successive layers of nutrient distribution. This would mean an adaptive response to nutrient depletion of increased sinking rates. Smayda (1970) suggests that the common occurrence of spines, protuberances and structured colonies represents adaptations enhancing exposure to nutrients because of the spiralling and looping that occurs due to these structures and shapes. While these considerations are appealing, the evidence so far is equivocal. *Ditylum brightwelli* shows changes in ionic composition which may represent an adaptive response of altered sinking (see above). Bienfang *et al.* (1982) have reported a direct test

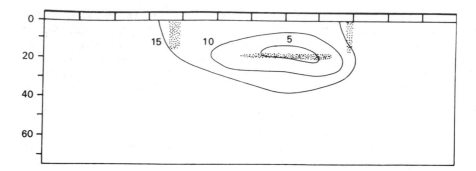

Fig. 3.6 Depth distribution of chlorophyll *a* maxima at station E1 in the English Channel in 1975 and 1976 (stippled areas), considered to represent the spring outburst, chlorophyll-rich thermocline layer and autumn bloom. The contours are an estimate of the stability of the water column, the values (in minutes) referring to a period calculated from the Brunt Väisälä frequency (*N*). Large values of *N* imply increased stability, the period in the contour lines is estimated as $2\pi/N$, so the smaller the period, the greater the stability. The greatest stability is therefore indicated for the depth around 20 m between June and September, in the region of the seasonal thermocline. (From Pingree *et al.*, 1977.)

of the hypothesis that nutrient impoverishment alters sinking rate. Working with four species of diatom and nitrate, phosphate and silicate, they found that depletion of silicate caused an increase of sinking rate in all four species. However, only in the case of *Coscinodiscus wailesii* was there an increase in sinking rate in response to depletion of nitrate and phosphate, while in *Skeletonema costatum* and *Ditylum brightwelli* nitrate depletion actually decreased sinking rate. Depletion of phosphate decreased sinking rate in these species and in *Chaetoceros gracilis*. Thus while nutrient impoverishment does indeed affect sinking rate, it is far from clear that this is an adaptive response to gain access to new nutrient supplies.

Dinoflagellate vertical distribution

As most dinoflagellates are motile, the problem of flotation is rather different for them. Swimming speeds have been measured, and these are of the order of 175–360 μm s^{-1} (Hand *et al.*, 1965) for *Gonyaulax polyedra* and a *Gyrodinium* species. In the case of *G. polyedra*, the observed swimming speeds are sufficient to account for the vertical migration undertaken by this form. Expressing the average speed of 250 μm s^{-1} in terms of daily swimming, a figure of 22 m d^{-1} is obtained. Sournia (1982) gives a table summarising work on dinoflagellate swimming speeds, which vary widely, the overall figure he derives is ~1 m h^{-1}. Sustained directional swimming is unlikely to be maintained for more than a few hours (see below), but these figures are useful for comparison with the estimates of diatom sinking speed and vertical circulation already discussed. The figure of 22 m d^{-1} is at the top end of the diatom sinking speed range, and similarly is small in comparison with vertical circulation estimates. Dinoflagellate swimming may be seen as offering an advance on sinking or rising (as used by diatoms and cyanobacteria) in influencing position in the water column, but it is still weak in comparison with even fairly modest turbulent mixing.

Dodge and Hart-Jones (1974) reported an extended survey of the occurrence and depth preferences of dinoflagellates in the North Sea, off the coast of Northumberland. The greatest abundance and diversity of dinoflagellates was in the summer and autumn, and the largest numbers occurred around 10 m. Different species showed preferences for different depths, with some fairly uniformly distributed through the water column. When dinoflagellates congregate in large numbers in layers near the surface, the result may be a red tide. These occur in calm weather, as mentioned above, but the subtleties of dinoflagellate behaviour involved in producing them are only just beginning to be understood. Cullen and Horrigan (1981) have shown that *Gymnodinium splendens* will migrate upwards to the surface during the day, but that in nitrogen deficiency this behaviour changes and the dinoflagellates remain at a depth corresponding to the 10% surface irradiance isolume. In contrast, *Gonyaulax polyedra* will maintain a surface distribution even in nitrogen shortage. It has been suggested that spending the night at 10–15 m depth allows cells to replenish their nitrogen supplies, after which they migrate to the surface for the day for active photosynthesis (Eppley *et al.*, 1968). Dinoflagellates are rich in carotenoids; as well as being involved in light harvesting these substances may have some role in protecting the pigment systems from photooxidative damage (Kirk, 1983), and this would be valuable in those species which form dense layers at or near the surface.

Phytoplankton cell size

The variation in size of phytoplankton cells has led to a tendency for classification into different size fractions. These divisions must be arbitrary, given the continuous nature of the size distribution, and the more there are the more burdensome does the list of names become. However it is probably justifiable and useful to distinguish between the nanoplankton (< 20 μm) and the netplankton or microplankton (> 20 μm). Most of the phytoplankton fall into the 2 to 200 μm range (using mean spherical diameter as a measure of size).

Nanoplankton are usually more abundant and make a greater contribution to total production than netplankton (Malone, 1980). This is especially true in oceanic waters, while the netplankton increase in coastal regions, including upwellings. However even in estuaries nanoplankton may be of overwhelming importance: McCarthy *et al.* (1974) considered that about 90% of primary production in Chesapeake Bay was attributable to cells passing a 35 μm mesh. This topic will be further discussed in Chapter 6. There are several possible reasons why cells of certain sizes might predominate at one time or another, and it is difficult to determine which may operate in any particular circumstances. It has sometimes seemed as if temperature was the single environmental variable with most influence, however as Sournia (1982) has commented, the data are contradictory. There does not seem to be any single environmental influence of supreme importance in governing cell size.

If certain sizes do have an advantage under different environmental conditions, then the relationship between cell size and physiological processes is likely to be involved in generating this advantage. It is important to be absolutely clear about the terms used in these discussions. As size increases, the rate of physiological processes increases, but it does not do so in proportion to size.

Using volume (V) as an index of size, the rate of a process (F) expressed as F organism^{-1} increases as about $V^{0.75}$, and expressing F and V as logarithms we can write:

$$\log F = a + \log V \times 0.75$$

where a = a constant (the intercept)
         0.75 = the regression coefficient in the equation

If the specific rate, that is the rate per unit mass or volume, is considered, in general the specific rate of metabolic processes is inversely related to size. The increase is now as about $V^{-0.25}$, and the regression coefficient similarly would be -0.25 for log specific metabolism on log volume. All this means that as size increases the absolute magnitude of processes increases more slowly than size, while the specific rate decreases. The study of allometric relationships of this sort is reviewed by Peters (1983).

As has already been mentioned (Chapter 2), there is an apparent relationship between the rate of nutrient uptake and cell size, with smaller cells showing a lowered value for K_s. This indicates greater uptake of nutrients at low steady-state concentrations by small cells than is possible for large cells, and this has been offered as a partial explanation of the tendency of small cells to dominate in oligotrophic oceanic conditions. Given the complexity of nutrient metabolism the relationship between K_s and cell size is of possible, but doubtful, significance. Friebele *et al.* (1978) have shown that the relationship between phosphate uptake and cell size in estuarine phytoplankton follows the pattern of allometric relationships described above. However the exponents calculated from their data are rather different from the norms quoted, namely 0.23 in the case of uptake rate on volume and -0.77 for specific uptake on volume (using logarithms in all cases).

Not only is there less chlorophyll *a* per unit volume in large cells than small ones (Malone, 1980), but also in large cells, chlorophyll *a* seems to be less active, as would be expected from the general allometric relationship between metabolism and size. Thus the light-saturated photosynthesis per unit chlorophyll *a*, or assimilation number (written P_{max}^{Chl}), may be influenced by cell size. P_{max}^{Chl} is usually two or three times higher in nanoplankton than in netplankton (Malone, 1980). In diatoms size-dependence of P_{max}^{Chl} has been demonstrated between 5 and 10 μm (mean spherical diameter), while from 25 to 170 μm it was independent of size (Malone, 1980). Photosynthesis is of course not the same as growth, since respiration takes a proportion of photosynthate, and respiration is also related to size in an allometric manner. Banse (1976) has reviewed the literature on algal metabolism and size. He used weight of carbon per cell as an index of size, and thus avoided the problems inherent in diameter or volume measures which arise from the presence of vacuoles in some species and not others. He concluded that both respiration and growth were size-dependent, with exponents sometimes less than unity in the case of growth but statistically indistinguishable from unity in the case of respiration. Given the similarity of the exponents, he concluded that growth efficiency (the proportion of photosynthate available for growth) was size-independent. Further work by Chan (1980) and Banse (1982) has confirmed the size-dependence of growth rates, though Banse (1982) considered it to be rather weak. Interestingly, different groups of algae may show different growth rates, and this will be referred to again below. Having discussed various aspects of size as such and of flotation and mixing, it is time to turn to succession and the composition of phytoplankton populations.

Phytoplankton succession

It has long been noted that there is something of a regular and predictable succession of phytoplankton following a spring bloom, or in the water moving from an upwelling. The distinction must be drawn between succession, in which different algae succeed one another in the same parcel of water, and sequential changes which may occur at one fixed place due to the passage of different water parcels each with its own flora.

Holligan and Harbour (1977) have described the succession in the English Channel. In spring and autumn, around the time of establishment and final decay of the seasonal thermocline, diatoms dominate. In early and late summer, there is a mixed flora of diatoms and dinoflagellates, while in mid-summer dinoflagellates and flagellates are dominant. This is the time of maximum water column stability (Fig. 3.6), and the dinoflagellates are found in the thermocline, with flagellates being more important in the surface layers. This pattern is probably typical of temperate coastal waters. Another interesting instance is the events accompanying the onset of the upwelling off Baja California, where dinoflagellates are important before the upwelling begins, with diatoms succeeding them (Walsh, *et al.*, 1974). This may represent a reverse succession, where disruption of a previously stable water column by the upwelling causes a reversion from dinoflagellates to diatoms.

The question of whether phytoplankton succession is directional and predictable has been addressed by Margalef (1958, 1967, 1978; see also Smayda, 1980). This is important in the context of the overall predictability (or otherwise) of planktonic systems, a topic which will be referred to again in Chapter 6. Margalef's view is of phytoplankton assemblages progressing from small cell species (principally diatoms) in nutrient-rich, turbulent water to larger cell species in more stratified water, and finally to flagellates under conditions of nutrient impoverishment. As things stand now, there is contradictory field evidence on these assertions – some studies are in accord with the idea of directional succession, others are not (Smayda, 1980).

Many factors have been proposed as being involved in driving phytoplankton succession. Smayda (1980) divided these into allogenic (those environmental conditions outside control by the organisms, such as salinity), and autogenic (those which may be regulated by the phytoplankton, or the zooplankton, including nutrients and predation). Rather different are sequential factors, which on the one hand may move populations by advection of water and on the other may alter the environment through turbulent disturbance.

In warm oligotrophic oceanic waters, which are comparatively non-seasonal environments, flagellates are generally dominant, with periodic diatom blooms punctuating this. In general terms, it is clear that turbulent conditions favour dominance by diatoms, and diatoms are the first to bloom as spring comes to temperate waters. They are succeeded by dinoflagellates and flagellates, or perhaps by cyanobacteria, depending upon location and conditions. Smayda (1980) suggested that in general motile species follow the initial dominance of non-motile ones. It seems that for substantial populations of flagellates and dinoflagellates to build up, a relatively stable water column is required. Part of the reason for this may be mechanical: it is said that flagellates and dinoflagellates are particularly susceptible to physical damage in turbulent water. One might be sceptical about this, since the organisms live in a low Reynolds number environment (see Chapter 5) where viscosity is relatively great. However there is observational evidence both from the sea and from cultures that strong turbulence does indeed cause cellular damage to phytoplankters. Another factor which may favour dinoflagellates (over diatoms) in stratified, stable waters is their ability to migrate down to nutrient-rich layers during the night, and up to the surface for photosynthesis during the day (see above).

Feeding by herbivores has been considered as a factor affecting succession. Selection on the basis of form, size and 'taste' might operate to change the species composition of the prey as a whole. Evidence that this occurs in the sea is hard to produce, unsurprisingly, given the other changes occurring at the same time. However Therriault and Platt (1978) have shown that grazing plays a role in the formation and maintainence of phytoplankton patches (see below). Also, lack of grazing has been inferred as a factor in the persistence of red tides when the dinoflagellate species are unpalatable (Fiedler, 1982).

Phytoplankton cell size in the sea

It is by now apparent that a variety of physical and biological factors influence cell size in phytoplankters. In diatoms successive divisions produce smaller and smaller cells until auxospore formation occurs, restoring cell size, but the matter that concerns us here is to do with the occurrence of populations dominated by cells of different sizes in different locations. We start with the observation that (successional changes aside) turbulent waters tend to be rich in diatoms, stratified ones in flagellates and dinoflagellates, and that coastal waters typically have phytoplankters with larger cells than are found in oceanic waters.

Semina (1972) has provided a synoptic picture of the distribution of mean cell diameters for phytoplankton in the Pacific (Fig. 3.7). Samples were taken with both bottles and nets, the bottle samples were relatively rich in small forms and these were found to be less variable across the whole ocean than were the netplankton. In the netplankton, three size categories were recognised: < 40 μm, 40–80 μm, and > 80 μm mean cell diameter. The average size at 518 stations in the Pacific was used to prepare an overall distribution of netplankton size categories, on which Fig. 3.7 is based. This shows a latitudinal shift in cell size, with smallest cells at highest latitudes. Along the equator there was a 'mosaic' of patches with large, intermediate or small cells predominating in different parts of the mosaic. It was apparent that large cells were found in the regions with persistent, stable ascent of water through the water column, determined as yearly averages. These regions straddled the equator, lying on either side of the main equatorial current systems. Where there was a tendency for water to sink, cell size was less, and it was least in those regions with comparatively vigorous vertical movement – at high latitudes and in the Peru upwelling. Semina (1972) concluded that cell size was influenced by the direction and velocity of water

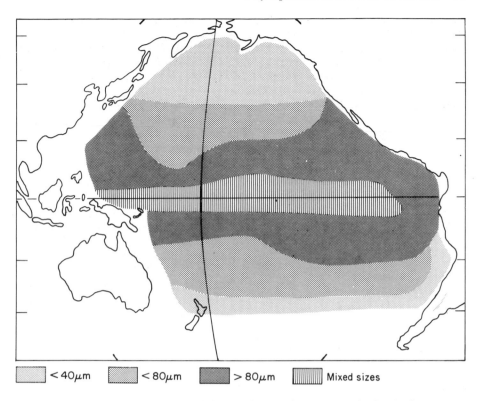

< 40μm	< 80μm	> 80μm	Mixed sizes

Fig. 3.7 Schematic representation of the distribution of cell sizes in the Pacific Ocean. For discussion, see text. (Redrawn from Semina, 1972.)

movements, the density gradient in the main pycnocline, and the level of phosphate. Vertical movements were seen as being most important, it was said that 'floating conditions' were best in the regions with the largest cells (which would tend to have the largest sinking velocities).

Parsons and Takahashi (1973) drew together information on physiological processes in two species, the relatively large *Ditylum brightwelli* and small *Coccolithus huxleyi*. They calculated growth rates under various conditions, and showed that the small species usually had a higher growth rate than the larger, except at the highest levels of light and nutrients in their experiments (Fig. 3.8). By extrapolation from this result, they suggested that differential growth rates might explain the observed cell size distributions in the sea. They developed an equation relating growth to light, nutrients, the depth of the mixed layer, and rates of sinking and upwelling, and used this to predict cell size under different conditions. Generally the picture was of small cells predominating in stable, oligotrophic waters, and large cells in more turbulent, eutrophic waters. This is somewhat at variance with the observations of Semina (1972), and Parsons and Takahashi (1973) point out that some of the regions characterised by Semina as having small cells (the Peru upwelling, the Antarctic) are known to have substantial biomasses of large cell phytoplankton as well.

The analysis of Parsons and Takahashi (1973, see also 1974) is interesting in

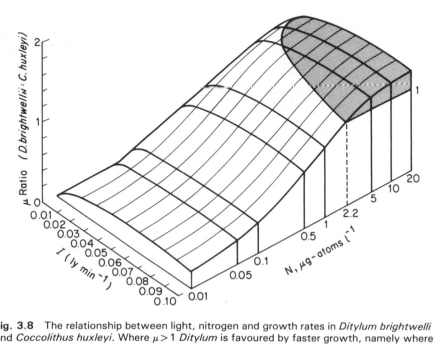

Fig. 3.8 The relationship between light, nitrogen and growth rates in *Ditylum brightwelli* and *Coccolithus huxleyi*. Where $\mu > 1$ *Ditylum* is favoured by faster growth, namely where both light and nitrogen are maximal. *I* is an estimate of the average photosynthetic light intensity in the water column. (From Parsons and Takahashi, 1973.)

that it seems, accurately, to predict the size of cell in natural phytoplankton assemblages, depending on a variety of conditions. Notably, large cells are predicted only under conditions of both high light intensity and high nutrient availability. Malone (1982) considered that large cells were selected under turbulent conditions when it may be important to last over unfavourable periods, using stored photosynthate. On the other hand in stratified conditions when nutrients may become impoverished, small cells are favoured because of their rapid nutrient uptake. These suggestions would explain the common predominance of nanoplankton, with microplankton assuming dominance only under special conditions. This work is the best current explanation of the relationship between phytoplankton cell size and environmental conditions, despite the doubts expressed by some workers (e.g. Friebele *et al.*, 1978).

Similar considerations may apply in the balance between dinoflagellates and diatoms. Chan (1980) has shown that diatoms have a higher growth rate than dinoflagellates, stemming from a higher photosynthetic capacity per unit biomass. This may in part explain the dominance of diatoms in turbulent and dinoflagellates in stratified water columns.

Patchiness

The sea is anything but homogeneous, and it has been recognised for some time that planktonic organisms are aggregated in patches of various sizes. This occurs in both the horizontal and vertical planes, and patchiness has what is called a length-scale (the physical size of the patch) and a time-scale (the longe-vity of the patch). Patchiness has recently been reviewed several times, one such paper is that of Platt and Denman (1980). They dealt specifically with the phyto-plankton, as we shall here, but much the same considerations apply to the zoo-plankton as well.

Circulation in the oceans occurs on a series of scales in time and space, from the major gyres which are relatively stable over years ($\sim 10^3$ km), to meso-scale eddies of the order of ~ 1 year ($\sim 10^2$ km), frontal eddies over weeks (~ 10 km), and smaller turbulent processes (such as breaking internal waves) over days or hours (~ 10 m or less). These time-scales and the temporal fluctuations imposed by seasonal events or such extrinsic disturbances as storms, interact with the bio-logical time-scale of population growth rates, producing patches of organisms on similar time- and space-scales. If population growth rate is shorter than the time-scale of turbulence then a patch has the opportunity to form, if it is longer then the growing population will be dispersed by turbulence. This qualitative argument is familiar from the consideration of the role of the critical depth in the timing of the spring bloom (Chapter 2). Denman and Platt (1976) propose that the length-scale of eddies be characterised by d, and a corresponding time-scale τ. This is the time taken for the eddies to transfer their kinetic energy to eddies of length-scale $d/2$. They then compare with a characteristic time-scale for phytoplankton, the reciprocal of the population growth rate r, namely r^{-1}. Then if $\tau << r^{-1}$ patches will not form, if $\tau >> r^{-1}$ they do. This simply pro-vides variables which, if they can be expressed quantitatively, allow a numerical expression of the previously qualitative argument. Attempts to do this and match the results with known patch sizes have produced a less than perfect fit, because other processes (such as grazing) are also acting on patches. Therriault and Platt (1978) report a study of St Margaret's Bay, Nova Scotia, in which they found evidence of patchiness on the scale of 1–5 km. However the situation was very dynamic, and the phytoplankton spatial distribution appeared to be con-trolled by a number of factors (including zooplankton grazing) of varying rela-tive importance. It was considered that patches did not persist for long enough to allow any one species to competitively displace others. This echoes a theme which has been discussed several times in terms of the persistence of many species in a fluid environment lacking solid physical barriers (e.g. Platt and Denman, 1980).

Actual patches of organisms are found in the sea with length-scales ranging over the orders of 10^{-2} to 10^3 metres at least. Larger patches are important as they represent aggregations of potential food for predators, and in such pheno-mena as red tides may have deleterious effects. Smaller patches may also repre-sent food for appropriately sized forms, and at the smallest scale patches of nutrients (rather than organisms) formed by zooplankton excretion may be important for phytoplankters. This has been mentioned in Chapter 2. The exis-tence of these tiny patches of nutrients is not in dispute, but their significance to phytoplankters in the sea is not yet established (Currie, 1984; Lehman and Scavia, 1984). Patchy food distribution and its importance for copepods will be discussed in Chapter 5.

4

Primary production and production cycles

Knowing something of the environment and its effects on phytoplankton, the major features of the cycles of production in various parts of the seas of the world can be more readily understood. It is convenient and appropriate to begin in the temperate mid-latitudes as these were the first systems to receive attention, and they remain the best known. The variability of production from site to site in one year, compounded by variability between years, makes interpretation of overall production levels very difficult. Nonetheless, because of the considerable interest in levels of primary production, values for annual net primary production will be given through this chapter. The most useful expression of this is the grams of carbon, fixed annually, under a square metre of sea surface, written $g \, C \, m^{-2} \, a^{-1}$, and this is the expression which will be used wherever possible.

This is not a book on methods and a thorough review of the methods underpinning the ideas discussed is beyond its scope. However it seems necessary to give a brief outline of the methods used in the estimation of primary production in the sea, so that the reader has an idea of how the data on this vital process have been obtained.

The estimation of primary production

There is a degree of constancy in the elemental composition of cytoplasm, for example the ratios of carbon:nitrogen:phosphorus atoms (C:N:P) in plankton are fairly constant at about 106:16:1 (Spencer, 1975). Moreover, the ratio N:P bears a close resemblance to the average ratio of these atoms in sea water. Using these relationships, early estimates of production were made by Atkins (1923) in the English Channel. Harvey et al. (1935) used similar methods. These workers measured the decline of phosphorus in the water column during the spring bloom, knowing the phosphorus content of the algae they were then able to estimate production of algal biomass. The method is liable to error if the assumptions about comparable ratios are invalid, and as phosphorus, in particular, tends to cycle and be re-used, this represents a further source of error. For the N:P ratio, values between 5:1 and 25:1 have been reported (Spencer, 1975). These reflect the tendency of nitrogen to vary with respect to phosphorus, the belief of early workers that these two elements might be taken up in strict proportion has not been substantiated. However an advantage of this sort of approach is that it measures changes of variables which occur as a result of processes in the water column.

An alternative is to investigate the changes in a sample of phytoplankton, enclosed in an experimental bottle, and illuminated. An early approach was to determine the oxygen produced as a result of photosynthesis. Pairs of experiments are set up, using clear and dark bottles. Oxygen evolution in the clear bottle is taken to represent net photosynthesis, while oxygen consumption in the dark bottle allows an estimate of respiratory losses – net photosynthesis plus respiratory loss gives gross photosynthesis. This assumes that respiration is unaffected by illumination or darkness, which is not true. This is not now a much used method.

The illuminated bottle approach has been extended using ^{14}C in a dilute $Na_2^{14}CO_3$ solution. The phytoplankton sample is inoculated with the labelled carbonate and illuminated, after this the sample is filtered and the radioactivity on the filter is measured. While this technique may bring a high level of precision to the estimation of ^{14}C incorporation, it requires the availability of good laboratory facilities. Again, precise or not, it is uncertain as to just what is being measured in terms of production – whether net or gross primary production, or some intermediate between these two extremes. Finally, this approach shares with the light and dark bottle technique the disadvantage that the experimental phytoplankton is enclosed in a highly artificial environment. Mixing is so important in constantly modulating light intensity and nutrient supply that the cessation of these effects in experiments may lead to very unrepresentative production estimates.

More recent approaches involve the estimation of changes in ATP in incubated samples, or refinement of the ^{14}C method in which the incorporated tracer is measured in a cell constituent (e.g. chlorophyll *a*) instead of in whole cells. Despite these developments (and sometimes because of them) there remains much uncertainty about true levels of primary production in tropical and subtropical waters, and this must be borne in mind in what follows.

Methods of estimating primary production are thoroughly discussed in Vollenweider (1969), for a more recent review of ^{14}C techniques in particular the reader is referred to Peterson (1980).

Seasonal patterns at mid-latitudes

In the North Sea, net primary production is negligible during winter due to the prevalence of mixing, combined with a shallow critical depth. As spring advances, the critical depth sinks through the water column which also begins to become stratified. At some time in this sequence of changes, the spring bloom commences and primary production increases. The details of the cycle will differ from year to year for the same location, as is seen in Fig. 4.1, in which production on the Fladen ground is illustrated. The Fladen ground stations lie in the North Sea off the Moray Firth, Scotland, extending about 100 km out to sea. Steele (1956, 1957) based most of his production estimates on depletion of phosphate, using a carbon-phosphorus ratio (see Chapter 2) to estimate carbon fixation from phosphate depletion. Experiments with ^{14}C gave results which were in good agreement with the ratio estimates. The estimates of primary production obtained by Steele for the North Sea varied between about 45 and 110 g C m^{-2} a^{-1}, with the Fladen ground value being about 68 g C m^{-2} a^{-1}.

Similar events, differing in detail, occur widely in temperate waters. One important variation is that as the water column becomes shallower the effect of

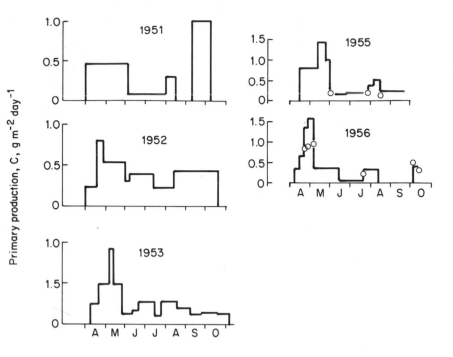

Fig. 4.1   Primary production on the Fladen ground in the North Sea. Open circles show rates measured by the ^{14}C method. (Redrawn from Steele, 1956, 1957.)

mixing in winter may be reduced, until in shallow estuaries growth may not be seriously limited by the winter reduction in radiation (Riley, 1967). Also in shallow water the spring bloom may be earlier in the year than it is in deeper water.

The initial peak in primary production is attributable to the response of phytoplankton populations to the onset of stratification in waters which are enriched with nutrients, increasingly well lit, and have small zooplankton populations. Pingree *et al.* (1976) identified the stabilisation of the water column, when the seasonal thermocline is established, as the crucial event triggering the spring bloom. The zooplankton numbers and biomass rapidly increase as the animals feed on the phytoplankton and reproduce, so that the spring bloom of phytoplankton is followed by a bloom of zooplankton (Fig. 4.2). After the initial phytoplankton bloom, numbers of phytoplankters and primary production typically fall during summer. This is attributable in part at least to zooplankton grazing, once populations of animals have increased, but nutrient shortage plays a role as well. The autumn bloom, sometimes evident at the time of the breakdown of stratification (Figs. 4.1, 4.2) suggests that the return of nutrients to the surface when mixing begins may have a beneficial effect on the phytoplankton, and is evidence of an effect of nutrient shortage during the summer.

Cushing (1975) has been particularly prominent in arguing that grazing by zooplankton controls the populations of phytoplankton after the initial bloom,

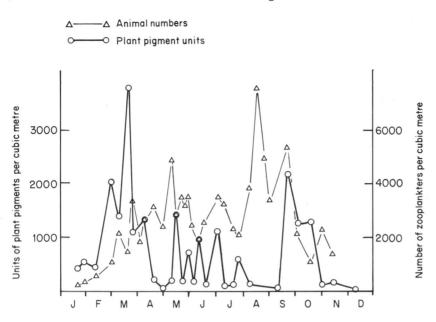

Fig. 4.2 Production at station E1 in the English Channel for 1934. Plant production was estimated in terms of units of plant pigment. (From Harvey *et al.*, 1935.)

thus depressing primary production below the potential level. He traces the recognition of the importance of the 'predator-prey model' to the work of Harvey *et al.* (1935), when it was appreciated that much of the algal net production was rapidly transferred to the zooplankton via grazing. Cushing (1975) considered that the 'predator-prey model' has replaced the 'agricultural model', in which primary production is seen as being nutrient-limited, save in special cases.

More recently attention has again been focused on nutrients and mixing, with the recognition of the importance of frontal zones in the sea. Pingree *et al.* (1976) considered that while the importance of grazing cannot be neglected, phytoplankton distribution in shelf sea areas such as the Celtic Sea (see Fig. 4.3) is much influenced by physical variables. It is particularly striking that at frontal zones between tidally mixed shallow water and deeper, more stratified water, there may be persistent blooms of phytoplankton (Fig. 4.3). The front in this instance seems to provide sufficient mixing for surface nutrients to remain relatively high, and at the same time sufficient stability for phytoplankton populations to develop. On either side of the front, conditions are less good: to the east because of vigorous tidal mixing, to the west because of more complete stratification. This evidence seems to provide at least partial vindication of the 'agricultural model'. Undoubtedly both grazing and nutrient limitation are potentially important in controlling phytoplankton production.

High-latitudes – the Polar seas

The seasonal pulse of production is more compressed than in temperate waters, there being a shorter period of the year when light levels are adequate for signi-

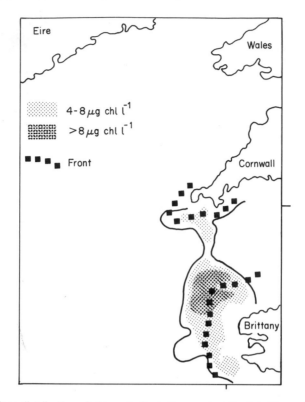

Fig. 4.3 Surface distribution of chlorophyll *a* in the western English Channel during the second fortnight of July 1975, together with the positions of fronts for the same period. The contour shows the outer bound of the ⩾2 *µ*g chlorophyll *a* 1⁻¹ zone. (Redrawn with modifications from Pingree *et al.*, 1976.)

ficant production. The pattern of production in the Arctic is relatively poorly known. A summary given by Nemoto and Harrison (1981) quoted values from 1 to 250 g C m⁻² a⁻¹, with most values lying between 50 and 100 g C m⁻² a⁻¹. Some of this variation is latitudinal: below 70°N production is higher than at higher latitudes, and under the ice, values are lowest (1–5) g C m⁻² a⁻¹ or lower). Low temperature, as such, is considered not to be important in limiting production, as the algae are adapted to the low temperatures. Indeed, significant production may occur in ice if there is sufficient light. Strong stratification (due to ice-melt) during the phytoplankton growing season first enhances conditions for growth, and then as nutrients are depleted acts to reduce production.

The Antarctic has traditionally been considered to be more productive than the Arctic, although there are widely differing values for production reported in the literature. Raymont (1980), quoting a variety of sources, gave values ranging from 20 to 130 g C m⁻² a⁻¹. Nemoto and Harrison (1981) gave similar values, but pointed out that some authors consider that much of the work aimed at estimating production in the Antarctic gives values which are far too high. It has been suggested that a realistic value is only 16 g C m⁻² a⁻¹. On the other hand, Jennings *et al.* (1984) have contradicted this low figure and restored the status

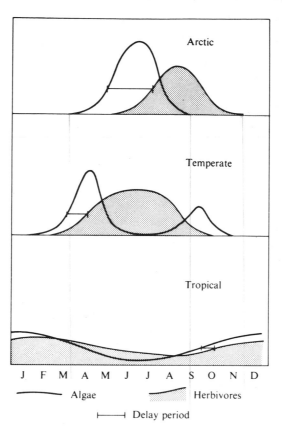

Fig. 4.4   A schematic representation of production cycles in different regions, emphasising the likely reduction of the lag between herbivore and producer populations with latitude. (From Cushing, 1975.)

quo with an estimate of the order of 100 g C m^{-2} a^{-1}. There is a marked latitudinal variation, with production being highest around the region between 70 and 75°S, under the influence of the upwelling, and lowest in the turbulence of the Antarctic Polar Front (Nemoto and Harrison, 1981). The most important factors controlling primary production are likely to be light, turbulence and grazing pressure.

The temporal sequence of production in high, mid and low latitudes has been summarised diagrammatically by Cushing (1975); his diagram is shown in Figure 4.4.

Upwellings

The physical basis of upwellings has been described in Chapter 1. Cushing (1971) has drawn an analogy between the temporal sequence of events in the temperate spring bloom, and the spatial sequence in upwelling. The source water for the upwelling is nutrient-enriched with respect to the surface, and it contains a sparse resident population of plants and animals. As the water moves upwards, it comes into the euphotic zone and the algae respond by increasing production and growth. With some degree of lag, the zooplankton respond to the increased availability of food, and their populations now increase. The increased grazing pressure in turn depresses the plant populations and diminishes further production as the water, now on the surface, moves away from the site of the upwelling. This simple model overlooks the extent to which both zooplankton and fish populations may use the current systems of coastal upwellings to maintain themselves within the system. Peterson *et al.* (1979) have given an account of the upwelling in the northern California Current off Oregon. This is a summer upwelling which alternates between vigorous and 'relaxed' states depending upon the winds, with a period of about eight days for a complete cycle. The water flow in a vigorous phase is shown in Fig. 4.5, also shown is the distribution of chlorophyll and major zooplankton groups according to Peterson *et al.* (1979). The chlorophyll maximum is not dissipated offshore because the seaward flow from the divergence is entrained in the southward 'jet' current. The authors proposed that the nearshore group of zooplankters are maintained there because they never encounter seaward flowing water. *Acartia longiremis* breeds in upwelling water and returns offshore at depths less than 10 m, while *Calanus marshallae* eggs and young are swept inshore of the divergence, to return seaward at times of relaxed upwelling when seaward flow of surface layers is more pronounced than is shown in Fig. 4.5. The overall effect is to produce a strongly-zoned distribution of organisms in successive offshore strips. How the populations are maintained in the face of strong southward flow is not known. While these distributions may be seen as the fortuitous outcome of the interaction of water movements and the behaviour patterns of the species concerned, the fact that coastal upwellings may retain zooplankters in this way means that they are more than spatial analogues of the temporal spring bloom.

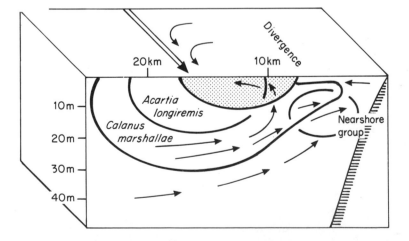

Fig. 4.5 Distribution of plankton during the vigorous state of the California Current upwelling off Oregon. The chlorophyll maximum is shown in stipple. The arrow on the surface represents the southward 'jet' current, which entrains seaward flow near the surface. For further explanation see text. (Drawn from information in Peterson *et al.*, 1979.)

Nevertheless, the idea that the biological significance of upwellings arises in part from the effects of water motion as such (and not only the injection of nutrients into the euphotic) is given support by Smith *et al.* (1983). They considered that in the Peru current upwelling, advective processes were the dominant forces controlling phytoplankton biomass.

From the viewpoint of the biologist, the main upwelling areas are those in the California, Peru, Canary and Benguela currents (see Fig. 1.2). There are also important upwellings in the Indian Ocean, occurring off the coasts of Somalia, Arabia and India. Occurring in the subtropics and tropics, these upwellings are responsible for marked changes in the plankton populations. These changes distinguish the plankton populations from those of the surrounding water, and the differences are reflected in the vertebrate populations. Walsh *et al.* (1974) have given a particularly useful account of the California Current upwelling. Off Baja California, the upwelling first begins in February, entering what the authors call the 'spin-up' phase in March, and becoming well developed in the summer. The pre-upwelling ecosystem in the surface waters is evidently based on the dinoflagellate *Gonyaulax polyedra*, grazed by herbivorous copepods such as *Acartia*. The copepods are preyed on by larvae of the red crab *Pleuroncodes planipes*, in turn preyed on by the Pacific striped dolphin, *Lagenorhyncus obliquidens*. Dinoflagellates at this time may be at an advantage because of their ability to migrate between the illuminated surface layers and deeper water, richer in nutrients. At its onset, the upwelling brings water from some 50 to 60 m, becoming progressively deeper until water is brought up from 100 to 200 m. This is accompanied by changes in the plankton: diatoms become dominant, the dominant copepod is now *Calanus helgolandicus*, and the red crab larvae become herbivorous, effectively shortening the food chain. The fish population changes to be dominated by tuna, feeding on the zooplankton and particularly the crab larvae. During the 'spin-up' period, production was about 7 g C m^{-2} d^{-1} over a fortnight, making carbon fixation up to nearly 100 g m^{-2} for this period alone.

Coastal upwellings last for part of the year – typical durations are 120 to 270 days, exceptionally 360 (Cushing, 1971). The relatively long duration of most upwellings, compared with the brief event of the spring bloom, is another factor in their overall importance. However upwellings are not necessarily constant, being stronger or weaker at times, and perhaps stopping altogether for periods. Moreover, at any one time upwelling will be more or less intense from place to place. Such a temporal fluctuation is the periodic cessation of the Peru Current upwelling, in the phenomenon known as El Niño.

In an El Niño event, anomalously warm and nutrient-poor water invades the surface layers off Peru and Ecuador. This occurs in the southern summer (i.e. in December through March) when the upwelling is weakest. Because of the major biological and commercial effects of El Niño, it has received a lot of recent attention. Barnett (1977) has considered the theories put forward to explain El Niño events. The simplest (the 'local wind' mechanism) is that local reduction of the longshore wind off the west coast of South America may lead to a cessation of upwelling because Ekman transport no longer moves surface water offshore. Alternatively, an imbalance between the southeast trades and the currents at the equator is supposed to be a potential cause, by allowing a flow of warm equatorial water southward (the 'overflow' mechanism). Or again, in the 'backflow' mechanism, it is envisaged that stronger than usual southeast trades build up water in the western Pacific over a period of about 2 years. This increases the slope of the sea surface from east to west, with the result that when the wind stress relaxes, there is a flood of warm water downslope, in the equatorial countercurrents, resulting in warming off the South American coast. In Barnett's analysis, this last mechanism emerged as the most important contributor to temperature changes in the eastern Pacific. However, El Niño is also clearly a complex phenomenon, and the events in the upwelling itself are part of a much larger sequence of events across the whole tropical Pacific (Philander, 1983).

The biological consequences are by no means uniform. The weak and brief event of 1975 reduced primary productivity in the equatorial ocean, but did not affect the fishery for the Peruvian anchoveta (Cowles *et al.*, 1977), because upwelling and high production continued adjacent to the coast. An intense event in 1972 did effect the fishery (Guillen and Calienes, 1981).

Two studies in the Peru Current upwelling may be used to illustrate the variability of conditions, even in the absence of a pronounced El Niño. Beers *et al.* (1971) found that in June 1969, the phytoplankton was dominated by flagellates and the zooplankton by ciliate protozoans. Carbon fixation was between 0.83 and 1.79 g C m^{-2} d^{-1}. Upwelling was relatively fast, at about 17 m d^{-1}. The authors considered that production was unlikely to be limited by nutrients or light, but did not exclude the possibility that organic 'conditioning compounds' might be limiting. They also ruled out grazing as a major limiting factor in this case, as apparently only about 25% of primary production was being grazed. A significant cause of loss of phytoplankton was considered to be by mixing of the upwelled water in the vigorous regime of the period. While upwelled water is often found at the surface in somewhat discrete patches, it appears that on this occasion such patches were rather short-lived, being rapidly mixed with surface water.

In contrast, Ryther *et al.* (1971) reported rather different conditions for a nearby area in the period March–April, 1966. They identified newly upwelled water by surface temperature measurements, whereby a fall from 20° to less than 16°C indicated the exact position of the upwelling. The upwelled water was tracked using buoys, and the changes in the plankton and nutrient levels were followed over five days. The phytoplankton was dominated by diatoms, and there was an increase of 50–60 g C m^{-2} during the five days of the study. About half of this had accumulated in the surface layer by the third day, but only about one fifth remained by the last day, with grazing supposedly having removed most of the balance. The authors were unable to investigate the zooplankton, as the phytoplankton was so dense that the nets clogged. However, the conditions they reported are those under which the Peruvian anchovy feeds directly on diatoms, thus shortening the food chain. In June 1969 the anchovy were feeding on zooplankton (Beers *et al.*, 1971), interposing one and perhaps two steps in the food chain in comparison with the situation reported by Ryther *et al.* (1971).

There is no doubt that primary production in coastal upwellings may be high. Some figures have already been given in the discussion above, but the matter is more than usually complicated because of variability between systems and between years within one system. Raymont (1980) gave values between 97 and

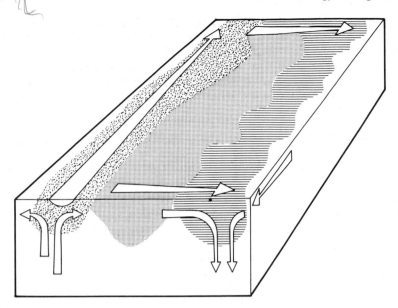

Fig. 4.6 Block diagram of the distribution of the trophic categories of plankton in an equatorial upwelling. The three zones indicated are of maximal abundances of phytoplankton, herbivorous zooplankton, and carnivorous zooplankton and fish, from left to right. Arrows show the direction of water movements. (From Vinogradov, 1981.)

190 g C m^{-2} a^{-1}, depending on latitude, as typical values for the Peru Current upwelling.

The events in equatorial current upwellings may be summarised as in Fig. 4.6. Upwelling tends to be more intense in the eastern than in the western ocean, as in the case of coastal upwellings. There is presumably the potential for the sort of retention of zooplankton that occurs in coastal upwellings, but this has not yet received much attention. Primary production may apparently be high, Vinogradov (1981) gave values ranging from 3.15 g C m^{-2} d^{-1} at 97° west in the eastern equatorial Pacific, declining westwards to 0.48 g C m^{-2} d^{-1} at 155° west, in January. It is not justifiable simply to multiply these figures for an annual estimate, but the evidence is that annual production must be high.

The oceanic gyral regions

In the tropics and subtropics away from the zones of coastal and equatorial upwelling and the boundary currents lie great areas of oceanic water which are relatively stable and undisturbed. These lie under the anticyclonic gyres in the atmosphere; these are large, stable permanent high pressure zones occupying the middles of the oceans. Distance from land precludes nutrient supply from land drainage, and the lack of the vertical movement of water associated with upwellings and mid- or high-latitude seasonal events allows no significant scope for nutrient renewal from below. It is in these waters that nitrogen limitation may be very important in controlling production, 'new' nitrogen being available only from rainfall, containing dissolved nitrogen oxides, and from fixation of elemental nitrogen by prokaryotes. Typically the open oceans are thought of as being very unproductive, but the figure of 78 g C m^{-2} a^{-1} given by Ryther and Yentsch (1958) for the northern Sargasso Sea is very much the same as the figures for the Fladen Ground in the North Sea (Steele, 1956; 1957). Admittedly, Ryther and Yentsch (1958) made a comparison of the Sargasso with shelf waters (100–160 g C m^{-2} a^{-1}) with Long Island Sound (389 g C m^{-2} a^{-1}), showing that in their study the oceanic water was far less productive. However, extending the comparison to Steele's work makes it clear that the various data on primary production must be interpreted with extreme caution. There is still no firm consensus on the levels of tropical and subtropical oceanic production. For example, Laws *et al.* (1984) and Sheldon (1984) argued that production may be higher than is usually considered to be the case, in oligotrophic regions. This will be referred to again in Chapter 6.

It is now known that the open oceans are influenced by events having their origin in the boundary currents. Cyclonic eddies form from the pinching off of meanders and then wander out across the open ocean, carrying a parcel of cool, nutrient-rich water known as a 'cold-core ring'. Such rings may have a life varying from several months to three years, and at any one time about 10% of the area of the northern Sargasso Sea may be covered by rings (Blackburn, 1981; Ring Group, 1981). Levels of chlorophyll *a* are much higher in rings than in surrounding oceanic water (Ring Group, 1981) and it is reasonable to suppose that primary production is similarly enhanced, although this does not have nearly the longevity of the ring (as a physical structure) as chlorophyll *a* levels are found to decline over the months. In terms of overall productivity, it is estimated that in the northern Sargasso Sea the regional productivity would be about 5% lower in the absence of rings (Ring Group, 1981). The ocean gyres are also subject to temporal disturbance of low frequency. McGowan and Hayward (1978) reported that the eastern portion of the north Pacific gyre underwent changes in the winter of 1968–1969, when sea surface temperatures fell and a negative temperature anomaly persisted for some time. This event was accompanied by an increase in biomass and productivity in the succeeding summer. Examination of oceanographic data was prompted by the biological response of increased production. The conclusions were that the breaking of internal waves nearer than usual to the base of the euphotic, together with overturn, produced by surface cooling, mixed an otherwise stable and stratified water column and enhanced production.

The overall picture

Despite the uncertainties involved in estimates of primary productivity in the sea, and the inconsistencies reported by the same authors between data obtained using different methods (e.g. see Ryther *et al.*, 1971), a pattern of regional variation in primary production is usually claimed for the seas as a whole. This is illustrated in Fig. 4.7, which reflects the variation in production levels already discussed, although there are differences in detail between the general pattern of Fig. 4.7 and some of the levels quoted above. Especially noticeable in the figure is the higher production evident all round the ocean margins, particularly in regions of upwelling. The equatorial upwelling belt in the Pacific is a prominent feature.

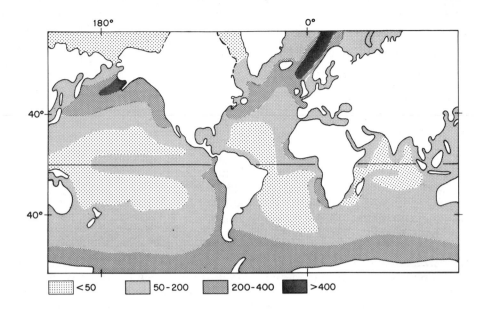

Fig. 4.7 Estimated primary production in the oceans as g C m^{-2} a^{-1}. (Redrawn from several sources.)

To an extent Fig. 4.7 reflects what may be called the classical picture of the distribution of marine planktonic productivity. There is no doubt that biomass does vary in the way shown in the figure, but there is uncertainty over production. The difficulties of measuring production by nanoplankton are such that there is wide disagreement between different workers, and as our knowledge of the planktonic food web increases it becomes less and less necessary to argue that production in so-called oligotrophic waters is really very low. This will be further discussed in Chapter 6.

As far as a global productivity sum is concerned, again data vary. The figures in Table 4.1 are taken from Lieth (1975), and show that marine primary production attributable to the plankton of the open ocean, upwellings and shelf seas, is rather less than half that of the continents as a whole. While the seas cover more than two thirds of the globe, they contribute just less than one third of global primary production (via the plankton). Even if we were to include an estimate of benthic production, from large algae and marine angiosperms, the total marine figure would still only come to 54.6×10^{12} kg a^{-1}.

Table 4.1 Estimates of global net primary production.

	Total area 10^6 km^2	Percentage	Total production 10^{12} kg a^{-1}	Percentage
Seas	359	70.7	50.9	29.5
Land	149	29.3	121.7	70.5

,e heterotrophic plankton

For reasons similar to those discussed in Chapter 3, this chapter is headed 'heterotrophic plankton'. Some of the forms to be discussed, although small and as yet poorly known, are unsatisfactorily classified as either animal or plant. A functional approach avoids this difficulty, but again it is not desirable to eliminate the familiar term, zooplankton.

The zooplankton are the animal portion of the plankton. The best known and in many respects most important are crustaceans, principally copepods. However, every major invertebrate phylum is represented, and as well as the holoplanktonic animals there is a host of larval forms of animals which are benthic or pelagic as adults. Many fish, including some important commercial species, have planktonic larvae. These larvae form the meroplankton.

Seasonal and vertical migration

It was seen in Chapter 3 that dinoflagellates in particular exercise control over vertical distribution in the water column, and indeed often show a diurnal vertical migration. In discussing upwellings, it was shown that the different depth distributions shown by developmental stages of certain zooplankters act to retain them within the coastal upwelling system off the Oregon coast. This may be termed ontogenetic migration, and is widespread in the zooplankton. Another example of this sort of pattern which maintains species distribution was discussed by Nemoto and Harrison (1981) for the Antarctic (Fig. 5.1). Krill (*Euphausia superba*) spend the summer at shallow depths south of the Antarctic divergence, drifting towards the continent. When they spawn the eggs sink to much greater depths, drifting north again. During development the larvae ascend towards the adult depth distribution. In other animals whole populations may show seasonal depth migrations leading to maintainance of horizontal distribution. The copepods *Calanoides acutus* and *Rhincalanus gigas*, and the chaetognaths *Sagitta marri* and *Eukrohnia hamata* spend the summer at relatively shallow depths north of the divergence, moving away from the continent in the surface water. After spending the winter deeper in the water column, they return towards the continent.

Of much shorter duration are the diurnal vertical migrations which are a common feature of the zooplankton. Characteristically the day is spent relatively deep in the water column, with animals swimming up to spend the night at shallower depths, perhaps at the surface in some species. Although this behaviour is widespread it is subject to considerable modification depending upon developmental stage and season. Figure 5.2 shows the distribution of different

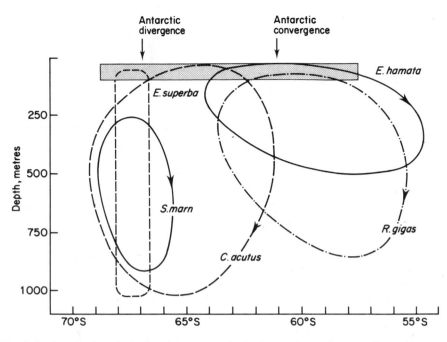

Fig. 5.1 Seasonal vertical migration patterns in the Antarctic zooplankters *Calanoides acutus* and *Rhincalanus gigas* (copepods), *Eukrohnia hamata* and *Sagitta marri* (chaetognaths), and *Euphausia superba* (krill). The shaded portion represents the summer feeding zone. (From Nemoto and Harrison, 1981.)

stages of *Calanus finmarchicus* in Loch Fyne in January and July. In January the females showed most evidence of migration, while the stage V copepodites (the immediate pre-adult stage) had a clear preference for greater depth than the adults. In July, females and stage IV copepodites showed active diurnal vertical migration, the stage V copepodites much less so.

The migrations shown in Fig. 5.2 are over about 100 m, greater vertical distances may be involved in some species. Omori (1974) has divided species of planktonic/pelagic decapod shrimps into seven groups, depending on their depth distributions, most showing vertical migrations over several hundreds of metres (Fig. 5.3). This figure shows two further important features of vertical migration: namely that it is a pattern of behaviour shown by animals at depth as well as near the surface, and that there is something of what has been called a 'ladder' of migrations reaching down into the ocean. This phenomenon seems likely to be important in the transfer of energy from the surface layers into the depths, although this is only beginning to be quantified. The greatest depth for which there is good evidence of diurnal migration is about 2000 m (Longhurst, 1976). This seems to place a constraint on the depths to which such a 'ladder' could operate.

Diurnal vertical migration is such a widespread habit, with so many variations of detail between and even within species, that it does not seem possible to provide a single all-important reason for its existence. For surface or near-surface forms, feeding in the rich surface phytoplankton at night will minimise visual

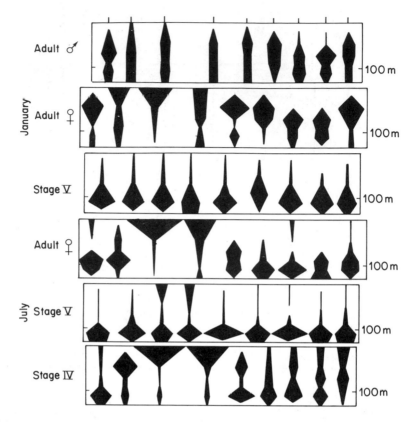

Fig. 5.2 Diurnal vertical migration indicated by changes in the distribution of *Calanus finmarchicus* in Loch Fyne, Scotland, in 1932. The distributions are for adult ♂, adult ♀ and stage V copepodites in January, and for adult ♀, stage V and stage IV copepodites in July. (Redrawn from Nicholls, 1933.)

predation, while spending the day in darker and cooler water may allow an energetic advantage due to reduced temperature, combined with an escape from visual predation. Intuitively it would seem as if the energy requirement for extensive vertical migration by small zooplankters might be prodigious, however this is evidently not the case. Vlymen (1970) concluded that for the copepod *Labidocera trispinosa*, the extra energy demand attributable to vertical migration was < 0.3% of the 'basal metabolic rate'. Vlymen's calculations have been criticised (see Lehman, 1977, and references therein) but his conclusions have received empirical support from the work of Foulds and Roff (1976). They found that in the freshwater mysid *Mysis relicta* swimming at rates equivalent to vertical migration speeds did not significantly increase oxygen demand over the 'routine' level. At higher speeds, they were able to show a 1.2 fold increase in oxygen demand. They concluded that there was evidence that vertical migration imposed a negligible cost in energetic terms. If this is so, then the reduction in metabolic rate resulting from spending the day in relatively cool water could be a distinct advantage for near-surface dwelling species. However, the results of Russian workers cited by Raymont (1983) suggest that the energy demand of

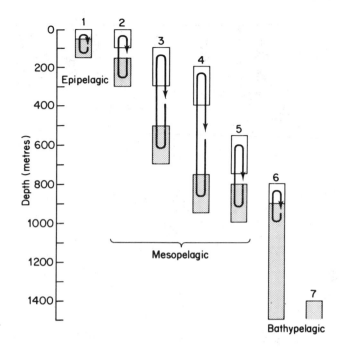

Fig. 5.3 Schematic representation of diurnal migration of pelagic shrimps. The main day and night concentrations are shown as stippled and hatched areas respectively. (From Omori, 1974.)

vertical migration may be higher than has been indicated: from 20–40% of basal metabolic rate, to 2 or 3 times this rate. The highest figure represents about 35% of the daily energy expenditure.

The most important factor controlling vertical migration is light. It seems to be the case that the animals are following an isolume; as it shifts up late in the day they follow it upwards, as it shifts down with dawn and daylight they sink or swim down. From the variety of actual responses it is clear that this is not a fixed behaviour pattern, but can be influenced by other factors. One such is the tendency of herbivorous zooplankton to aggregate at subsurface chlorophyll maxima, this modifies their vertical migration excursions. Here, the habit of vertical migration and the ubiquitous patchiness of phytoplankton interact. Mullin and Brooks (1976) have shown that in the case of *Calanus pacificus* the heterogeneous distribution of algal food means that at many places in the surface layers the copepods are apparently unable to meet their food demands, while at some they will. Clearly, success will depend in part on finding and then remaining in suitable food patches. In a comparison of the effects of starvation on copepods, Dagg (1977) has shown that *Acartia tonsa* and *Centropages typicus* were comparatively sensitive to food shortage, dying in 3–10 days without food. *Pseudocalanus minutus* and *Calanus finmàrchicus* on the other hand survived > 16 days and > 3 weeks, respectively, without food. Egg production was also differentially affected, being depressed by lack of food in *Acartia* and *Centropages* but not in *Pseudocalanus* (*Calanus* did not produce eggs in the

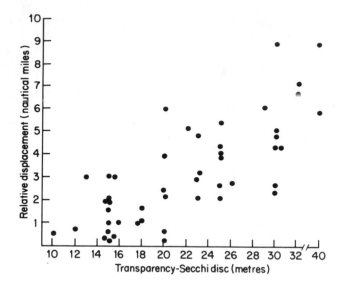

Fig. 5.4 Estimated daily displacement of deep scattering layer organisms relative to the surface plotted against water transparency (Secchi disc depth). (From Isaacs *et al.*, 1974.)

experiments). It may be concluded that food patchiness in space or time will differentially affect copepods.

Isaacs *et al.* (1974) have shown that the daytime depth of the deep scattering layer in the eastern Pacific is correlated with the transparency of the water column. This is consistent with the idea that the animals track an isolume; the clearer the water, the deeper a given isolume will be, and thus the deeper a given group of animals will be. Transparency of the water is largely dependent upon the amount of phytoplankton, thus clear water is relatively poor in food. When relative current speeds and directions with depth at the sampling locations were taken into account, the authors found that there was an excellent relationship between transparency and displacement relative to the surface (Fig. 5.4). This occurred because of the fact that currents at different depths flow in different directions. Thus, greater vertical migration, as happened in clear water, had the effect of removing animals relatively rapidly from clear, food-poor surface regions, while the converse would tend to retain them in rich areas. This work confirms and extends ideas originating with Hardy (1953) concerning the relationships between vertical migration, food requirements and patch formation.

Heterotroph feeding

This topic has exercised marine biologists for several decades. In terms of the economy of the sea, feeding by the herbivores is enormously important, for it is the process whereby much of the net primary production of the phytoplankton is harvested, then to be incorporated into animal tissue and made available to larger animals. The mechanics of this process, and the quantities involved in successive transfers, are of great interest and importance. By far the most important zooplankters numerically are the calanoid copepods, and they have received most attention.

Particle capture by calanoid copepods

Calanoids have usually been regarded as filter feeders, which is understandable in view of the form of their mouthparts. There are eleven pairs of appendages along the body, the first five considered as being on the head, namely the antennules, antennae, mandibles, and first and second maxillae. The rest are on the thorax: the maxillipeds, and five pairs of swimming limbs or pereiopods. In higher crustaceans there is fundamentally the same arrangement, although

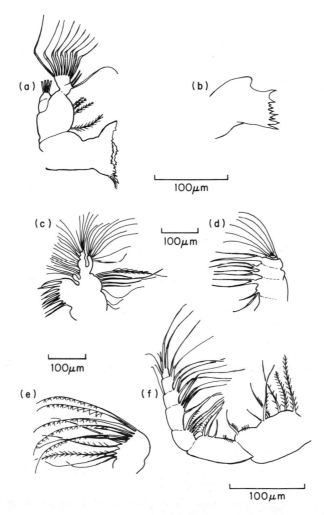

Fig. 5.5 Mouthparts of calanoid copepods. **(a)** Mandible of *Paracalanus aculeatus*; **(b)** mandibular blade of *Acartia spinata*; **(c)** first maxilla of *P. aculeatus*; **(d)** second maxilla of *Temora turbinata*; **(e)** second maxilla of *A. spinata*; **(f)** maxilliped of *P. aculeatus*.

terminology differs slightly: the first maxilla may be called the maxillule, followed by the maxilla. This is followed by more pairs of limbs than in cala-noids, for example, in euphausiids there may be between six and eight pairs of thoracic limbs, depending on genus, followed by six pairs of abdominal limbs. The first five of these are called pleopods, followed by one pair of uropods, forming part of the tail fan.

Figure 5.5 shows examples of the mandibles, maxillae and maxillipeds of calanoid copepods. The crustacean limb is basically biramous, consisting of portions which in copepods are termed the basipodite, exopodite and endo-podite. This is best seen in the mandible of *Paracalanus aculeatus* in Fig. 5.5. The blade is borne on the basipodite. The other limbs shown have been so modi-fied that the original form is obscure, or has been changed by loss of one or other of the exopodite or endopodite. *Paracalanus* is a common genus of planktonic copepod. For comparison, the blade of the mandible of another small inshore form is shown, namely *Acartia spinata*. The first maxilla characteristically bears many fine, plumose setae and a set of stouter setae on the basipodite, pointing inwards – this is well shown in the example of *Temora turbinata*. The second maxilla bears many plumose setae as well, and sometimes there is one stout one – again shown in *Temora*. The second maxilla of *Acartia* has unusually long setae, associated with its method of feeding (see below). The maxilliped, formed from the first thoracic limb, again is equipped with long, plumose setae.

Feeding by calanoids was first studied by Esterly (1916). In his account of feeding in *Eucalanus elongatus* and *Calanus finmarchicus*, studied with sus-pensions of carmine particles, the animals were seen to gather particles into the mid-ventral line and direct them forwards to the mouth between the setae of the second maxillae. The actual gathering was achieved in swirling water currents generated in an unspecified way. At the mouth, the particles were formed into a pellet which was sometimes ingested (by *Calanus*), and sometimes broken up and dispersed (always, in the case of *Eucalanus*). Esterly (1916) pointed out that the animals from the field often had diatom remains in their guts, and he con-sidered that they were able to filter sufficient particles from the phytoplankton to satisfy their requirements (although perhaps this was possible only in dense phytoplankton patches). This conclusion was important, for there was, for many years, a body of opinion that there was insufficient particulate food in the sea for copepods, and that they must be meeting at least part of their nutritional requirements by uptake of dissolved organic matter. This view was held parti-cularly by the physiologist Pütter, advanced in papers published in German in the first quarter of this century. A useful review is that by Jørgensen (1976).

Cannon (1928) gave an account of the water movements produced by *C. finmarchicus* held in a volume of water just large enough to allow unhindered movements of the limbs, and observed with a microscope and stroboscopic light. He described a pair of large 'swimming vortices' associated with the slow forward swimming of the animal, and a pair of smaller 'feeding vortices' com-plementary to them, which passed water forward through the mouthparts (Fig. 5.6). These vortices arose as a result of the rapid beating of the antennae, mandibular palps and first maxillae. The second maxillae did not beat, but were mostly held still in the water flow, and it was supposed that these formed the filter. Figure 5.7 shows the second maxilla of *Calanus* with three algal cells of differing sizes for comparison, this drawing epitomises the large body of work

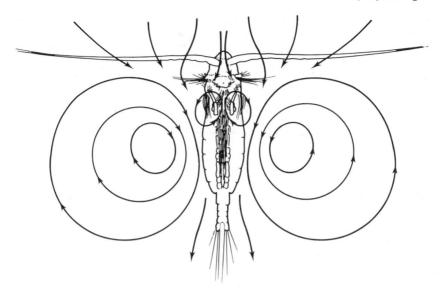

Fig. 5.6 The 'feeding currents' described for *Calanus finmarchicus*, now thought to be artefactual. (From Cannon, 1928.)

in which the second maxilla is seen as a filter. Captured particles were said to be removed by the setae of the first maxillae, and passed forward to the mandibles.

Both Esterly (1916) and Cannon (1928) had pointed out that particles were sometimes vigorously rejected. However there was a tendency to regard calanoid feeding as being 'automatic', a byproduct of the animal's swimming. This was reinforced by the results of the first quantitative experiments on the rate of particle removal, which indicated that food particles were taken in proportion to their abundance over the experimental range (Fuller, 1937; Gauld, 1951). Therefore if the process was seen as one of filtration, the volume of water being filtered remained constant, while the ingestion of particles rose with particle density (Fig. 5.8a). However, more recently it has become clear that ingestion does not necessarily increase with particle density (Frost, 1972), and moreover that below a lower threshold of particle density feeding may cease (Frost, 1975). This modified understanding is shown in Fig. 5.8b. As copepods are capable of considerable plasticity in their ingestion rate and are able to track particle density over a wide range with increased ingestion, the so called 'upper critical concentration' may only rarely (if ever) be encountered in the sea (Conover, 1978). There is indeed the possibility that the upper thresholds demonstrated in experiments do not reflect natural behaviour, as the experimental animals may have been exposed to unusually high cell concentrations for abnormally long periods (Parsons *et al.*, 1967; Conover, 1978).

That feeding cannot be 'automatic' is shown by the discrimination exercised over which particles are ingested. Harvey (1937) showed that calanoids could appreciate the nature of the particles they encountered, and this has been confirmed by several recent workers. For example, Poulet and Marsot (1978) showed that of two sorts of synthetic particles, those including an extract from

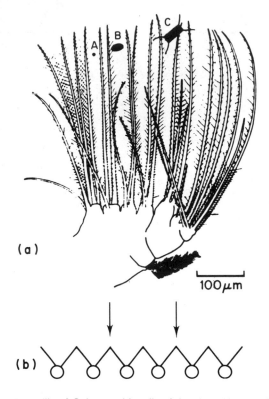

(a)

100μm

(b)

Fig. 5.7 (a) Second maxilla of *Calanus* with cells of the algae *Nannochloris oculata* (A), *Sycracosphaera elongata* (B), and *Chaetoceros decipiens* (C) drawn to the same scale. (b) Representation of how the setules on the setae might act in filtering, the setae are seen in cross section with the setules spread across the current (arrows). (From Marshall and Orr (1960) In: *The physiology of Crustacea, Vol. I*, T.F. Waterman, (ed.). Academic Press, New York.)

natural phytoplankton were selected for ingestion. Friedman and Strickler (1975) described chemoreceptors on the mouthparts of the fresh water calanoid *Diaptomus pallidus*. Further evidence for behavioural complexity comes from the response of copepods to bioluminescent dinoflagellates. Esaias and Curl (1972) and White (1979) have shown that *Acartia* and *Calanus* reduce their ingestion of bioluminescent cells at high levels of bioluminescent activity, suggesting that the animals are 'startled' by the flash of a cell stimulated by their appendages. Altogether there is considerable evidence for selectivity in calanoid feeding.

Apart from the study of filtering, it has been known for some time that various sorts of grasping behaviour are involved in calanoid feeding. In *Acartia* the first maxillae are relatively large, and they may be used in what has been called 'seining': the setae are spread, and then quickly drawn together, capturing suspended particles (Conover, 1956). Gauld (1964) suggested that this sort of particle capturing method may be quite widespread in the important calanoids of the plankton, including *Calanus*.

These observations have recently been extended by Koehl and Strickler (1981)

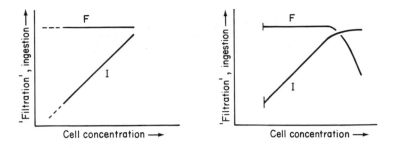

Fig. 5.8 Diagrams summarising development of ideas on 'filtering' in copepods.
(a) constancy of filtration (F) leading to increase of ingestion (I) with cell density;
(b) filtration (and therefore ingestion) cease below a lower cell concentration, while above
an upper critical concentration filtration decreases and ingestion remains constant.

and Paffenhöfer, *et al.* (1982) using copepods between 2 and 3 mm long. Koehl
and Strickler (1981) used high speed cinematography to study the movements of
the appendages of *Eucalanus pileatus* and *Centropages typicus*, following the
movement of algal particles near the animals and tracing water flow with
streams of Indian ink. The animals were tethered with a short hair cemented to
the thorax, this did not show any sign of interfering with their behaviour. The
volume of water in the observational cuvette was 120 ml, very much larger than
the small drops of water in which previous observations had been made (e.g.
Cannon, 1928). Koehl and Strickler (1981) suggested that the vortices described
by Cannon (1928) were an artifact produced by enclosing the animal in a very
small volume of water. They found that water was not pumped through the
second maxillae, but rather that the movements of the antennae, mandibular
palps, first maxillae and maxillipeds produced a pulsed flow of water past the
animal. With an algal cell in the vicinity, the beat became asymmetrical so as to
draw water preferentially from the direction of the cell. Finally the second maxillae were flung apart, causing water to rush between them, carrying the algal
cell. The maxillae were then closed over the captured water parcel and cell. As
this happened, the water round the cell was forced between the setae of the
second maxillae to escape posteriorly. The captured particle was now transferred to the mouth using the short endite setae of the first maxilla to comb the
setae of the second maxilla.

Both water and particle movement ceased immediately the motion of appendages ceased – there was no sign of inertial movement of either water or
particles. The activities described are shown in Fig. 5.9. It is noteworthy that the
Indian ink flowed as undisturbed streams, suggestive of laminar, non-turbulent
flow. Were turbulence to be generated by the copepod's appendages this would
have dispersed the ink into the surrounding water.

Koehl and Strickler (1981) went on to consider the physical aspects of fluid
behaviour as they apply to copepods, using the dimensionless expression of the
ratio of inertial to viscous forces known as the Reynolds number (Re). This is
calculated from:

$$Re = \frac{\rho \, v \, L}{\mu}$$

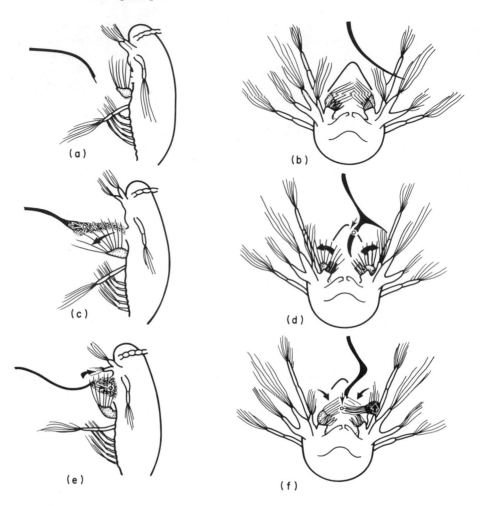

Fig. 5.9   Feeding in *Eucalanus pileatus*, these drawings were traced from high-speed cinematographs of tethered individuals, and show events as seen from the side (**a, c, e**) or front (**b, d, f**) of the animal. The black streams indicate India ink, the heavy black arrows appendage movements, and the fine arrows movements of a cell (open circle). Currents bypass the second maxillae (**a, b**) until a food particle is near when the second maxillae fling out (**c, d**) and sweep in (**e, f**) capturing the particle. In (**f**) India ink is shown escaping between the setae of the right second maxilla, it is apparently only at this stage that water flows between the setae of the appendages which otherwise are acting as paddles rather than filters. (From Koehl and Strickler, 1981.)

where ρ = the density
          μ = the dynamic viscosity of the fluid
          v = the relative velocity of the fluid across a solid object of linear
              dimension L

When Re is low, flow is laminar, but turbulent when Re is high. Using the film records Koehl and Strickler (1981) were able to estimate velocities of limbs and flow rates, knowing limb dimensions and the viscosity of the sea water they were able to calculate Re, and the values were found to be very low (10^{-2} to 10^{-1}). This indicated laminar, non-turbulent flow, which was consistent with the observations described above. The conclusion the authors came to was that for most of the time, the appendages were acting as paddles rather than as open rakes or sieves.

Using similar techniques, Paffenhöfer *et al.* (1982) have confirmed these findings for *Eucalanus pileatus* amd for *E. crassus*. For particles > 10 μm diameter or length, feeding appeared to be by grasping. These authors do leave open the possibility of another method of feeding, unspecified but said to be 'closer to a purely passive filtering motion', for capturing very small (< 5 μm diameter) particles.

This recent work has been given particular attention here, because it persuasively argues that filtering, if it occurs at all in calanoids, must be much less important than many authors have supposed to be the case. Thus while Boyd (1976) and Frost (1977) considered that the experimental facts concerning particle removal rate supported the idea of passive mechanical filtration, conclusions drawn from recent observational work and the consideration of fluid dynamics are to the contrary. This work suggests that the raptorial feeding often described in copepods, and considered to be used for relatively large particles, in fact may apply down to particle sizes of about 10 μm. Down to this sort of size, the animal captures a parcel of water containing the particle, and then expresses the water through the setae. This is referred to as the 'scan and trap mechanism' by LaBarbera (1984). The behaviour of the animals suggests that they can appreciate the presence and nature of particles several hundred diameters from the appendages, this may be accomplished by sensing chemical or mechanical clues. Chemicals diffusing out from cells will form a shell of 'odour' around the particle'which may be detected by chemoreceptors such as have been shown to exist by Friedman and Strickler (1975). In a low Reynolds number environment the drag on one object may be changed by another at some distance, and this allows the possibility of mechanical appreciation of particles before contact is made (LaBarbera, 1984).

Particle capture by mysids and euphausiids

Much less attention has been paid to zooplankters other than copepods. Among the Crustacea, mysids and euphausiids are important zooplankters besides the copepods. They show a variety of techniques for feeding on suspended particles, for resuspending bottom deposits and feeding from the resulting clouds of detritus, or for acting as predators.

The feeding current in these animals is produced by the exopodites of the thoracic limbs, which are held out from the body and move so as to whirl around the axis of the limb ramus, drawing a current of water down the ramus and into the mid-ventral line (Fig. 5.10). Here there is an open channel between the limb bases, the food groove, in which water and suspended material is drawn forward by the beating of the mouthparts (Cannon and Manton, 1927; Mauchline and Fisher, 1969). Particles are said to be filtered out on the setae of the maxillae (= second maxillae) and first thoracic appendages, and passed forward to the mandibles. Mysids are small (usually < 15 mm) but larger than copepods, and

(a)

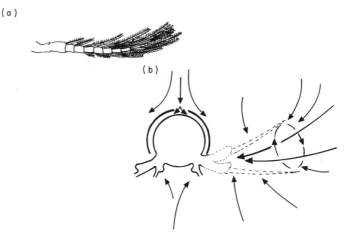

(b)

Fig. 5.10 The thoracic exopodite **(a)** and water currents generated by their movements **(b)** in *Hemimysis lamornae*. In **(b)** the thorax is seen as in section from the front, one exopodite is shown and the direction of beat indicated. (From Cannon and Manton, 1927.)

euphausiids are often larger still, it is more feasible that true filtration may be involved in particle capture among these animals than among copepods. At the same time there seems to be no reason why they should not be using 'scan and trap' processes just as do copepods.

The euphausiid *Meganyctiphanes norvegica* is capable of feeding when the exopodites of the thoracic limbs have been removed, suggesting that the mouth-parts alone produce sufficient water flow for feeding, and that the exopodites are important in bringing water from around the animal to the food groove (Mauchline and Fisher, 1969). Euphausiids may also resuspend bottom material using a current of water generated by the pleopods, after which the animal backs away and takes particles from the suspension. Alternatively, material may be ploughed up from the bottom with the antennae, the animal then rises slightly and spreads the thoracic limbs, suspending the material in a space called the food basket, formed between the thoracic limb bases and the mouthparts. Large particles are then masticated by the mandibles. Figure 5.11 represents these activities. A similar spreading of the thoracic limbs, causing an inrush of water, may be used to capture animal prey by drawing it into the food basket (Mauchline and Fisher, 1969). Similar habits are reported for mysids (Mauchline, 1980).

Particle capture by other heterotrophs

There is a comparative dearth of experimental information on non-crustacean zooplankter feeding. The varied forms to be mentioned here share the characteristic that they feed wholly or in part on very tiny particles.

For some time it has been clear that the hosts of tiny, non-pigmented flagellates in the plankton may be important grazers of bacteria. They represent a taxonomically heterogeneous assemblage of forms, some apparently related to protozoa and others to dinoflagellates and euglenoids. Fenchel (1982) outlined the structural features of some important forms, two of which are shown in

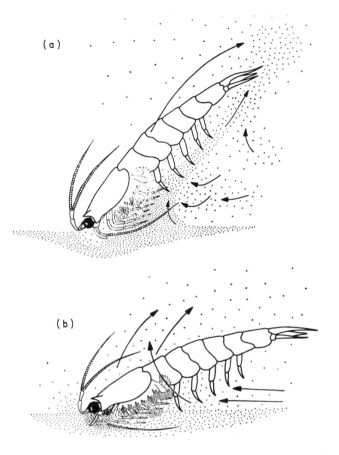

Fig. 5.11   Euphausiids feeding on bottom detritus by **(a)** lifting it into suspension using water currents generated by the pleopods (abdominal swimming legs), or **(b)** by ploughing up material while moving slowly along the bottom. (From Mauchline and Fisher, 1969.)

Fig. 5.12. A water current is produced by the flagellum (in the biflagellate forms only one flagellum is active in producing a current). If the cell is free, the current drives the cell through the water. In the case of *Pleuromonas*, bacteria are intercepted by the cell and ingested via the cytopharynx. In *Actinomonas*, there is a cone of pseudopodia which is said to act as a filter or sieve. The details of the feeding processes remain obscure. According to Fenchel, many of the forms spend much of their time anchored to suspended particles, when their flagella may function more efficiently in feeding.

   An important group of larger forms is the larvaceans, such as *Oikopleura*. These animals are primitive chordates, somewhat resembling tadpoles in shape. The body consists of a pear-shaped head (or trunk) with a long, ribbon-like tail. The animal lives in a 'house', a hollow gelatinous structure provided with channels and filters suspended within the house. A pair of incurrent channels, guarded by coarse filters, lead into the house cavity and fuse to form a chamber in which the animal lies (Fig. 5.13). This chamber divides near the tip of the tail,

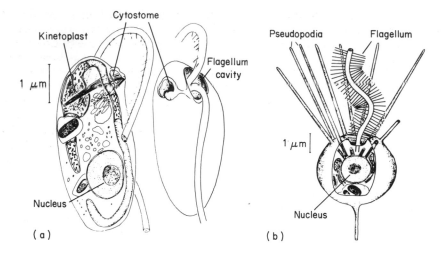

Fig. 5.12 The heterotrophic flagellates **(a)** *Pleuromonas jaculans* and **(b)** *Actinomonas mirabilis*. (After Fenchel, 1982a.)

each branch leading into the feeding filter. The beating of the tail pumps water down the incurrent channels past the animal and then into the feeding filter. This consists of dorsal and ventral layers which are corrugated so that the structure resembles a partially boxed duvet, the water flows in between these layers and passes out through the fine meshes, collecting in the space outside the filter

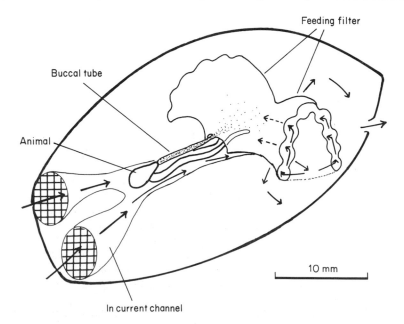

Fig. 5.13 The larvacean *Megalocercus huxleyi* in its house. Arrows show the directions of water currents, the operation of the filter is described in the text. (After Alldredge, 1977.)

before emerging from the house via the exit pore. Particles remain suspended in a concentrated residual flow within the filter, this concentrated suspension is sucked towards the mouth along the buccal tube. Once in the pharynx the suspension is filtered by being drawn through the pharyngeal mucus net, and the mucus and trapped food are ingested. This account of feeding is based on the work of Alldredge (1977) and Deibel (1986).

Periodically the filters become clogged, when this happens the animal abandons the house and builds a new one. The longevity of a house is reported to be between four and six hours. Alldredge (1976) has studied larvaceans *in situ* using scuba diving. Observations of living animals made by divers have been increasingly used recently, and have greatly extended our understanding of feeding behaviour of many planktonic animals. Alldredge (1976) found that the animals secreted a new house as soon as feeding in the existing one began, and were reluctant to abandon the house until the filters became clogged. Secretion and expansion of a house were completed remarkably quickly. Secretion itself might take more than 10 minutes, but expansion of a secreted house took only 1.5 to 5 minutes.

Paffenhöfer (1973) cultured *Oikopleura dioica* through numerous generations on algae 4–5 μm in diameter. He considered that next to the copepods, larvaceans are the most numerous zooplankters, and very important in view of their ability to feed on very small phytoplankters. Their ecological importance is enhanced by their ability to feed on bacterioplankton (King *et al.*, 1980) which may supply 12 to 50% of their daily requirements.

We also know something about feeding in a numerous group of pelagic opisthobranch gastropods collectively referred to as 'pteropods'. There is disagreement over whether these belong to one order (Pteropoda) or two (Thecosomata and Gymnosomata) within the Opisthobranchia. At any rate, they are pelagic 'snails' which often lack a shell, and in which the foot has been expanded into two lateral parapodial wings. These are used in active swimming; it is therefore a moot point as to whether these are strictly planktonic animals in the sense of 'drifters'. The Gymnosomata are carnivores; little is known about them. In the Thecosomata, the wings are used both in swimming and in food collection, by the gathering of fine particles. Yonge (1926) described the feeding of four species, of which *Cavolinia inflexa* was the simplest (Fig. 5.14). On the posterior margins of the wings is a ciliated field, the Wimperfeld, histologically a lateral extension of the middle lobe of the foot. These fields are covered with mucus gland cells and ciliated cells; particles become trapped in the mucus and are driven towards the lobes of the foot by the cilia. Here they are collected, the animal evidently exercising selectivity before ingestion. Above the mouth, in a groove between the paired lateral lobes, is a rejection tract for carrying away unwanted material. Considering all four species, Yonge (1926) described a progressive reduction in the area of the ciliated fields until in *Gleba cordata* they are reduced to a pair of ciliated tracts, one on each side of the animal. This is accompanied by a reduction in the structures of the buccal mass, inside the mouth. *Cavolinia* is equipped with jaws, radula and salivary glands, all of which are absent in *Gleba*.

Morton (1954) studied *Limacina retroversa*, even less advanced than *Cavolinia*. He drew attention to a large glandular region in the mantle cavity, the pallial gland, present in all thecosomes. The pallial gland is also mucus

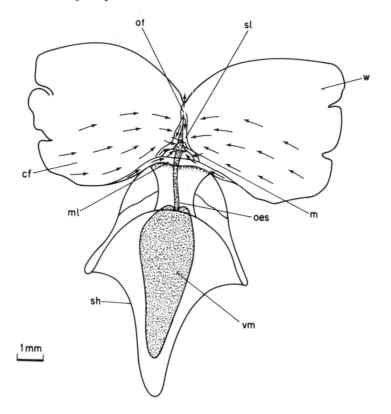

Fig. 5.14 The pteropod mollusc *Cavolinia inflexa*: cf, ciliated field or Wimperfeld; m, mouth; ml, mid lobe of foot; oes, oesophagus; ot, ciliated rejection tract; sh, shell; sl, side lobe of foot; vm, visceral mass; w, wing. (From Yonge, 1926.)

secreting, the mucus trapping particles in the water current swept through the mantle cavity by the inhalant cilia on the left rim of the mantle. Larger particles (of the size of diatoms) are trapped in the mucus, the mass being formed into a string and carried forward on the right floor of the mantle cavity, thence out of the mantle cavity and to the mouth (Fig. 5.15). The radula is used to grip the food string for ingestion. The foot also collects and directs particles to the mouth, however the cilia of the Wimperfeld direct particles away from the mouth, and these are apparently not used as food.

Gilmer (1972, 1974) studied feeding in eleven thecosomatous pteropods, including four species of *Cavolinia*, using direct underwater observation by scuba diving as well as laboratory studies. Gilmer (1974) found that the pallial gland and mantle cavity currents are important in *Cavolinia*, as they are in *Limacina*. Indeed the sequence of events seems to be that food is trapped in mucus in the mantle cavity, formed into food strings and extruded from the mantle cavity to form a suspended mass of food strings. These are subsequently collected by the cilia of the foot lobes and Wimperfelds and passed to the mouth. Periodically the animals might break away from the suspended mucus strings and swim away to begin feeding again. Gilmer found that particles in the

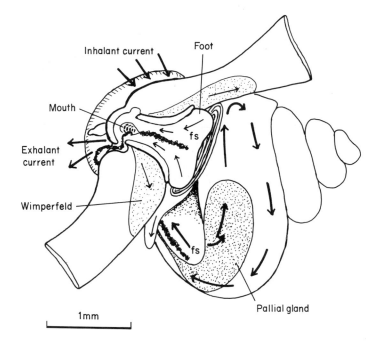

Fig. 5.15 The pteropod *Limacina retroversa*. Thick arrows show the direction of water currents into and through the large mantle cavity, leaving on the animal's right. Thin arrows show the direction of ciliary beat on the epidermis of the foot and Wimperfelds: fs, food string. (After Morton, 1954.)

size category < 10 μm were the most important numerically, forming between 20 and 30% by number in the food strings examined.

The most remarkable findings concern the advanced forms, *Gleba cordata* and the related *Corolla spectabilis*. Yonge (1926) described the extreme reduction of the ciliated tracts in *Gleba*. Gilmer (1972) has shown that these are not used in the same way as in other pteropods, rather, both *Gleba* and *Corolla* produce relatively enormous webs of mucus which extend as a free-floating, unsupported sheet of mucus fibres of 1–6 μm width. The fibres criss-cross, to leave minute pores; the area of these pores varies from about 500–3500 μm², 500 μm² being the most common size category. The animal is very small (~ 50 mm in the case of *Gleba*) compared with the size of the web, which is up to 2 m in diameter. The feeding animal hangs immobile and upside down under the web, with the mouth in contact with the mucus at the end of the extended proboscis. Both web and animal sink slowly through the water. Particles are trapped on the web, both in the meshes and by adhesion, and the cilia of the tracts on the proboscis pull particles towards the mouth. There they are consolidated into a fine mucus bound food string and ingested. At the approach of danger (such as the observer at a distance of about 1 m) the animal rapidly detached from the web, righted itself and swam away. Sometimes, an animal might detach to swim to another portion of the web and begin feeding again. Figure 5.16 is drawn from a photograph in Gilmer (1972), and shows *Corolla spectabilis* swimming.

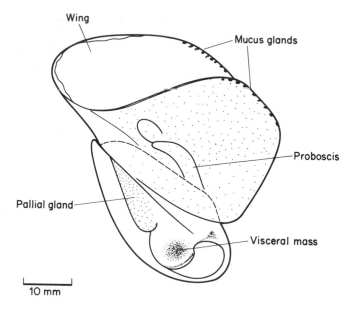

Fig. 5.16   The pteropod *Corolla spectabilis* swimming towards the surface of a tank in the laboratory. The large mucus glands on the periphery of the wings secrete the web (not shown here) which acts as a large free-floating food trap. (Drawn from a photograph in Gilmer, 1972.)

Sources of food for calanoids

Whatever the details of the method of capture, it is now thought that in fact calanoids do meet their food requirements by ingesting particulate food, and do not need recourse to the uptake of dissolved organic substances. While dissolved organic substances are known to be taken up by a variety of marine invertebrates, and may have a role in the energy budget, this has not been recorded in the Crustacea (Stewart, 1979). The apparent uptake of dissolved amino acids by some Crustacea has been shown to be due to the activity of microorganisms on the exoskeleton (Jørgensen, 1976). This is in contrast to the situation in fresh water, where there is evidence of the uptake and utilisation of dissolved glucose by copepods (Gyllenberg and Lundqvist, 1978). It is clear also from the above review that many other planktonic animals have adaptations for capturing and feeding on suspended particles. Even the heterotrophic flagellates, often supposed to subsist on dissolved organic matter, seem more likely to be feeding as particulate feeders (Fenchel, 1982). While they can undoubtedly take up dissolved organic matter, the concentration of this in the sea seems to be too dilute for it to be used directly by the heterotrophic flagellates. Rather, it is taken up by bacteria which are then grazed by the flagellates.

Animals such as *Calanus finmarchicus* are omnivores, feeding on phytoplankters and on other very small crustacea, as well as on detritus (Corner *et al.*, 1974). The omnivorous habit is likely to be widespread among small calanoids. All members of the phytoplankton appear to be used as food by copepods except for the cyanobacteria, which are generally not favoured. An exception to this is the harpacticoid *Macrosetella gracilis*, which feeds on *Trichodesmium* and indeed requires it in the diet (Roman, 1978). When phytoplankters are plentiful, copepod guts are packed with algal cells and are often green in colour. This is one of the observations which led some workers to question how efficiently the food was being assimilated, raising the problem of 'superfluous feeding', discussed below. At mid- and high-latitudes, phytoplankton will not be abundant all year, and at times of shortage (as during winter) other foods become more important. *Calanus finmarchicus* will feed on living and dead nauplii, and in experiments can obtain greater amounts of nitrogen and phosphorus from this source than it does from phytoplankton in the spring bloom (Corner, *et al.*, 1974). These authors suggested that natural animal food was necessary to sustain overwintering *Calanus*. Natural particulate material from the mouth of the Clyde in Scotland, collected during winter, was inadequate. So was particulate material produced by foaming sea water in the laboratory. On both of these diets, the animals lost substantial amounts of body nitrogen and phosphorus during the course of the experiments (20–30% over five days), while a diet of nauplii allowed them to maintain body nitrogen and phosphorus. Similar conclusions were reached by Heinle *et al.* (1977) for estuarine copepods. In contrast, Poulet (1976) has shown that *Psuedocalanus minutus* in the inshore waters of Nova Scotia, consistently ingested a higher proportion of non-living than living carbon. It appeared in this instance that detritus was a main source of food, not an incidental one.

Among larger calanoids, and particularly those living in deeper water, carnivory becomes more important. *Labidocera* and *Pontellopsis* are effective predators not only of other copepods but also of young fish larvae in the California Current (Lillelund and Lasker, 1971). The copepods are attracted by the vibrations from the tail of a larva in the vicinity, swimming towards the larva and grasping it with the mouthparts preparatory to ingestion.

Superfluous feeding

Beklemishev (1962) proposed that while copepods respond to increased food availability by increasing ingestion, they do not necessarily maintain the same assimilation efficiency as ingestion increases. There might therefore be a phenomenon which he called 'superfluous feeding' in which the faeces of the animals become progressively richer in undigested algal matter. This would obviously be of potential importance for animals on the sea floor, especially in shallow water, as a 'rain' of rich faecal pellets might be expected under conditions of superfluous feeding. Beklemishev (1962) considered that such conditions might be quite widespread after the spring bloom, or in upwelling areas immediately after upwelling. However there is no experimental evidence from work with marine crustaceans which would support the inverse relationship, between ingestion and assimilation, necessary for the existence of superfluous feeding. Among marine copepods, the assimilation efficiency is usually at least 50% and may be > 90% of the organic matter in the food (see Grahame, 1983 for review). There is no evidence of a consistent decline in assimilation with increase in ration.

It is interesting that, on the one hand, an apparent paucity of particulate food led A. Pütter to conclude that the direct uptake of dissolved organic matter must be nutritionally important, while on the other, conditions in the spring bloom have led to suggestions of superfluous feeding. In the sea, copepods will encounter rapid and often unpredictable changes in food availability, and rapid adaptations take place in the animals to allow them to cope with fluctuating food supply (Conover, 1978; Mayzaud and Poulet, 1978). They are evidently able to ingest large quantities (up to about 100% of body weight per day), assimilating most of this and using it in metabolism, and for growth, storage and reproduction (Gaudy, 1974). Under conditions of abundant food, reproduction becomes a major 'sink' of assimilated carbon, at least in laboratory experiments.

Carnivory

As has been mentioned above, a great many calanoids are to some extent carnivorous. But the plankton also contains specialised carnivores, and important among these are the arrow worms, or Chaetognatha. These form a small phylum, among invertebrates, of 52 species. They are marine or estuarine, and all are carnivorous, a few are benthic, the majority are planktonic. It is the planktonic chaetognaths with which we are concerned; in the plankton they are often next in abundance to the copepods (Feigenbaum and Maris, 1984). In comparison with surface dwelling planktonic copepods they are relatively large, of the order of 10–30 mm. Prey are detected by sensing vibrations, the animals will attack a glass or metal probe if it is vibrated in the water. They therefore depend upon the presence of live, active prey, which must be fairly close to the animal (2–3 mm) for detection. Young chaetognaths feed on such animals as ciliates and nauplii. Among adults the most important food items are copepods, making up between 32 and 99% of the prey by numbers (Feigenbaum and Maris, 1974).

There are different opinions concerning the ecological importance of chaetognaths in the plankton. A summary of estimates of their consumption, as a percentage of secondary production, gives values ranging from 2.2 to > 100% (Feigenbaum and Maris, 1984). The highest value refers to *Sagitta elegans* in St. Margaret's Bay, Nova Scotia, in winter. At this time chaetognath biomass was high and the calculated energy demand of the population exceeded copepod production. However it was considered that in terms of the fate of the total copepod production through the year, chaetognaths were relatively unimportant, taking only ~ 1% of the annual production (Sameoto, 1972). In contrast in the adjacent Bedford Basin, chaetognaths were estimated to take 36% of annual copepod production (Sameoto, 1973). Grahame (1976) inferred that chaetognaths were important in regulating copepod numbers in a eutrophic tropical harbour. However given that most of the studies of these animals are for inshore waters, leaving their oceanic importance largely unexplored, it is still true that we do not know what the significance of the chaetognaths is in the seas as a whole.

6

Planktonic food webs

This chapter deals with the fate of the carbon fixed in algal photosynthesis, and aims to show how important is a consideration of food chains to understanding planktonic systems and the potential for fish yields. Because the various necessary terms have been used in a variety of contexts, and often to mean different things, it seems desirable to begin with some definitions.

Terms and concepts

Elton ([1927], 1966) used the term food chain to refer to the sequence of animals linked by feeding one upon the other, ultimately sustained by plants at the beginning of the chain. He referred to all the food chains in a community as the food cycle, this term has been replaced by the expression food web. Food chains are interlocked into a pattern of trophic relationships since animals almost always have a wide variety of foods and predators. Elton also drew attention to the progressive decrease in numbers up the food chain – herbivores were more numerous than the carnivores which preyed on them, and secondary carnivores were less numerous again. This he called the pyramid of numbers.

Implicit in the idea of a food chain is the organisation of living things into trophic levels. Thus green plants belong in the primary producer level, they may be termed photoautotrophs. Chemoautotrophs are also primary producers, however they are unimportant in the plankton. The herbivores are primary consumers, feeding on the plants. First carnivores, feeding on the herbivores, are secondary consumers, and so on. At each trophic level, there is production. This is most obvious in terms of the first level, where photosynthesis results in primary production. Much of this will be respired by the producers themselves, and so we can distinguish between gross (or total) and net (after respiration) primary production. Net primary production is sometimes called the apparent photosynthesis. It is this (or rather a fraction of it) which may be available to the next trophic level. In turn, the organisms of each consumer level ingest food, assimilate a fraction of it (defaecating the rest), respire much of the assimilate, and incorporate some as new tissue – this represents production at that trophic level. Thus the consumers are responsible for secondary production, which by definition is a net amount as it can only be considered to have taken place after respiration. Odum (1983) referred to all production at consumer levels as 'secondary production'.

At each level some energy is lost in excretion; in the recommended nomenclature of the International Biological Program this is not regarded as having been assimilated, however this conflicts with common usage among marine biologists (see Grahame, 1983, for discussion). Commonly, assimilation is taken to be ingestion minus faeces. Whether nitrogenous excreta are regarded as part of the assimilated or not, they are certainly excluded from the production value for the trophic level.

Energy is the ability to do work, for organisms food energy is available as the chemical energy of high energy bonds in various substances. Photoautotrophs initiate this store of compounds when they synthesise carbohydrates using light energy. This is an example of the transformation of energy from one form to another, however energy can neither be created nor destroyed (first law of thermodynamics). No process of energy transformation will spontaneously occur unless there is a degradation of some of the energy from a concentrated to a dispersed form (second law of thermodynamics). Commonly this dispersed form is heat; once energy has been dissipated as heat in biochemical processes, it is unavailable for work. There is substantial loss of energy at each trophic level, the energy available for the next trophic level is then only a fraction of that available to the present level. Some effort has been directed at finding if there is any pattern in this. A ratio

$$\frac{\text{Energy passed to } n + 1}{\text{Ingestion at } n}$$

may be defined where n is the trophic level. Expressed as a percentage, this is called the ecological or food chain efficiency (Slobodkin, 1962), and it has been suggested that it will have a value of between 5 and 20%. This means that in the plankton, herbivorous copepods might be expected to make available between 5 and 20% of their ingested food to their predators.

The distinction between the biomass at any one time, or the standing stock, and the rate of increment of this biomass (productivity), is a crucial one. In Elton's pyramid of numbers, the numbers generally decrease up the pyramid (from producer to successive consumer levels), and the same is true if numbers are converted to biomass as standing stock (Odum, 1983). However, inverted or partially inverted pyramids of biomass (or numbers) are sometimes found, and one example of this is in plankton food webs. An often quoted example is for the English Channel, where 4 g m^{-2} dry weight of primary producers sustained 21 g m^{-2} dry weight of consumers (Odum, 1983, quoting Harvey, 1950). This was said to be possible because of the high ratio of productivity to biomass (P/B) obtaining in the primary producer trophic level. In general, the smaller the organism, the higher this ratio will be. While pyramids of numbers or biomass may be inverted, those of energy flow cannot be: the amount of energy available to successive trophic levels must decrease, consistent with the second law of thermodynamics. Examples of biomass pyramids are shown in Fig. 6.1, taken from the work of Holligan *et al.* (1984a) on English Channel plankton. These authors distinguished between three water column regimes: stratified (E5), frontal (F) and tidally mixed (M). Only in the first of these was there an inverted biomass pyramid. The authors were cautious in interpreting the inverted pyramid for station E5, saying that the factors allowing the persistence of this situation are not precisely understood. It would of course be possible to construct inverted pyramids of energy flow from these data; this does not

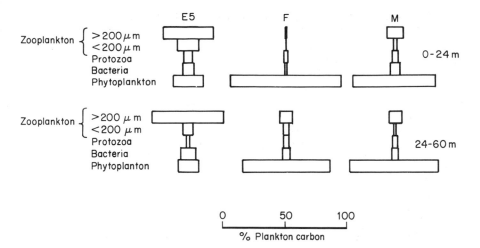

Fig. 6.1 Percentage of plankton carbon in 5 categories of organisms in the plankton of stratified (E5), frontal (F) and tidally mixed (M) water columns in the English Channel in 'summer' (the last week of July, 1981). The thermocline at E5 extended from about 18 m to 25 m, with a weaker thermocline between 15 m and 28 m at F. The dominance by heterotrophs at E5 possibly indicates relatively high efficiency of utilisation of plant carbon in stratified water, with little export to benthic communities. (From Holligan *et al.*, 1984a.)

indicate the reality of such a system, but rather the difficulty of extrapolating from such information.

It is axiomatic that once the chemical energy contained in high energy bonds has been used in respiration, it is unavailable for further respiration. This consideration imposes a fundamental distinction between energy and materials in ecosystems, for while materials can cycle and the system may be at least partly closed in this respect, energy is constantly dissipated in ecosystems. Therefore in terms of energy, an ecosystem must be open – it depends on constant input. Rigler (1975) maintained that since some photosynthate may escape from primary producers into solution, then to be consumed by other organisms, it can be said that energy also cycles. This seems to miss the point: what may be said to be cycling are the molecules in which the energy is trapped until used, once used it is gone.

Before moving on, the difficulties arising from attempts to classify organisms into trophic levels should be considered. Rigler (1975) has objected to the whole notion of trophic levels on the grounds that (with the exception of primary producers) it is impossible to unambiguously place organisms in one level or the other. This is because of the fact that most consumers will feed partly in two or more of any possible scheme of trophic levels. Considering the example of the food web of the herring (*Clupea harengus*) determined by Hardy (1924), Rigler queried any attempt to assign it to a particular trophic level. Figure 6.2 shows the percentage contribution of various foods to the diet of the herring throughout the year, weighted for differences in prey size and the numbers of herring feeding. It can be seen that 47% of the food comes from the sand eel *Ammodytes* and the chaetognath *Sagitta*, themselves zooplankton feeders. Another 18% comes from the hyperiid amphipods, which again are carnivores

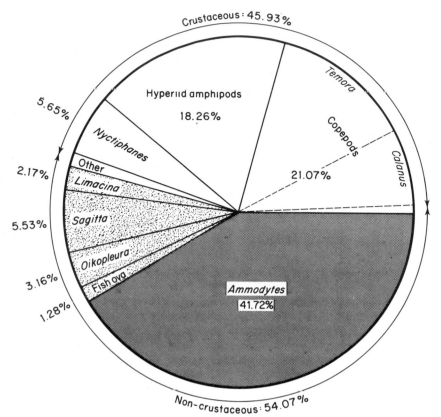

Fig. 6.2 The relative importance of the main food categories taken by herring (*Clupea harengus*) throughout the year. The diagram is divided into portions reflecting the estimated contributions by weight of the different categories, taking into account seasonal changes in food availability and feeding behaviour through the year. The category 'other' includes a wide range of very occasional foods such as barnacle and decapod larvae, Cladocera, and unspecified 'foods of minor importance'. (From Hardy (1924) who notes that representing the figures for percentages to the second decimal place gives a spurious precision to the estimates.)

(sometimes on herring larvae; see Sheader and Evans, 1975). Therefore for 65% of its food, the herring has at least two trophic levels between itself and the primary producers, so it is here a tertiary consumer. For 32% of its food, the herring is nearer the phytoplankton with only one trophic level interposed, acting as a secondary consumer.

Hardy (1924) also compiled a food web for the herring, shown in Fig. 6.3. Up to a size of 12 mm, the young fish feed on small food including diatoms and other phytoplankton. At larger sizes phytoplankton disappear from the diet and copepods (*Pseudocalanus* and *Temora*) become very important. The adult herring feed on zooplankton and juvenile sand eels (*Ammodytes*). The information represented in Fig. 6.3 is complementary to that in Fig. 6.2, and is valuable in showing how the food requirements of the fish change through life as it

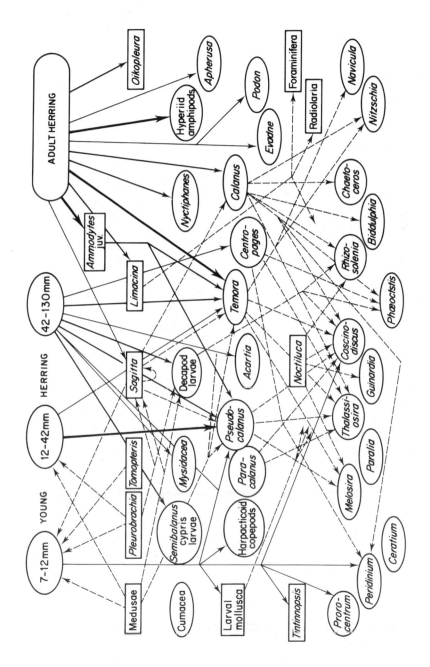

Fig. 6.3 The food web of the herring (*Clupea harengus*). The direction of the arrows indicates the source of the food and not the transfer of energy. Solid lines from the work of the original author, broken lines from the work of others. (From Hardy, 1924.)

grows, and in giving an impression of the complex network of relationships in the food web. This is an example of a food web for a single organism feeding fairly high in the food chain.

Clearly it is not realistic firmly to categorise the herring as a herbivore or carnivore (it is both, at different stages of its life), nor to assign it to one trophic level only. This kind of difficulty is common, and it led Rigler (1975) to reject, completely, the idea of trophic levels as being useless, maintaining that it is unproductive of testable hypotheses. As shall be seen, this is not the case, but the ambiguity of trophic classifications must always be borne in mind.

The fate of fixed carbon

Extracellular products

A sizeable fraction of phytoplankton photosynthate escapes from the cells into the sea as soluble carbohydrate such as glycollic acid ($CH_2OH. COOH$). Fogg (1966) suggested that 5–35% of photosynthate might be lost, depending on conditions. There is doubt over whether the 'excretion' of glycollate is an active or a passive process (Harris, 1980). It apparently may occur when growth is limited by nutrients (as opposed to being light-limited), and photosynthesis is proceeding faster than the rate at which new carbohydrate may be incorporated into structural material or respired, or it may occur when cells are physiologically stressed (e.g. in photoinhibition). Sharp (1977) queried whether healthy cells lost significant amounts of photosynthate to the sea, but agreed that the loss of soluble products from senescent cells or those disrupted by zooplankton feeding might be considerable. He suggested that phytoplankters might not be able to afford to lose quantities of photosynthate in this way; however this argument overlooks the possibility that glycollate excretion is in part a 'safety valve' for voiding excess photosynthate.

Berman and Holm-Hansen (1974) found that in eutrophic waters in the tropical Pacific, between 6 and 12% of fixed carbon was lost as extracellular products, in oligotrophic waters this rose to 17–27%. They noted that these levels of loss were lower than has sometimes been reported, and suggested that experimental error had contributed to what they considered to be previous overestimates.

Fogg (1966) suggested that the principal beneficiaries of the excretion of soluble extracellular products might be bacteria. He pointed out that as these substances do not accumulate in the sea, they must be being consumed in some way. The extracellular products such as glycollic acid must be distinguished from much of the bulk of total dissolved organic matter in the sea, which is said to be highly resistant to microbial use (Jørgensen, 1976). This total is estimated to amount to about 1.5×10^{18} g in the whole oceans, about 30 years worth of primary production. The dynamics of the refractory portion are very poorly known, but more is now being learned about the pathways taken by molecules such as glycollates. They seem to be of little or no direct account to the larger heterotrophic plankton, but the suggestions that they are important to bacteria have received ample support. For example, Bell (1983) studied the transfer of organic carbon from algae in culture to marine bacteria from the water column. There was rapid transfer of organic carbon from the algae to the bacteria, via an extracellular products pool. This was considered to occur as a result of the ability of bacteria to use very dilute substrates.

The transfer of organic matter from phytoplankton to bacteria via the pool of dissolved material, and the importance of regeneration of nutrients in the euphotic in sustaining further production, both present a challenge to the idea that most of the net primary production is harvested by zooplankton her-bivores. This view has been put by Cushing (1975) and also by Steele (1974), among others. More recently the microheterotrophs have been seen to be of great importance, and the consequences of this for our understanding of plank-tonic food webs will now be considered.

Examples of food webs

There is a great deal of fragmentary information on what planktonic animals eat at different times and places, but there is a distinct dearth of thorough studies embracing even a modest subset of the organisms of any given plankton community. This is particularly true if quantitative information is sought. The reason is not hard to see: such studies are difficult to do because of the enormous effort required. Again, it seems desirable to begin in inshore temperate waters and extend understanding from there.

Harvey (1950) produced a study of the production and fate of organic carbon in the sea off Plymouth, based on his observations and those of others, over a period of years. One of the principal conclusions was that about half of the net primary production was assimilated by planktonic herbivores, the remainder being taken by benthic animals. These, together with the demersal fish feeding on them, were the ultimate beneficiaries of planktonic production.

The waters of the English Channel and North Sea have recently been the object of more study with a view to further elucidating the pelagic food webs, using microbiological techniques unavailable to earlier workers. In the English Channel, advantage has also been taken of better understanding of the structure of the water column. Holligan *et al.* (1984b) have investigated the primary production and heterotrophic energy demand at three stations in the western English Channel: in a frontal region (F), and in tidally mixed (M) and seasonally stratified (E5) water to the east and west of the front, respectively. These are the same stations as were referred to above in the discussion of pyramids of biomass and energy flow. Phytoplankton biomass was highest at stations F and M, and so was primary production. A striking feature of all three stations was that consumption by microheterotrophs (here including bacteria, protozoans and microzooplankton) appeared to account for most of the available production. The respiration of microheterotrophs was greater than that of mesozoo-plankton by factors of 9, 43 and 17 at stations E5, F and M respectively. The authors remarked that phytoplankton production and microheterotroph consumption were in close balance with one another.

Using data from the same three stations, but for August, Newell and Linley (1984) described the likely pathways of carbon flow. A simplified version of their representation of the food webs is shown in Fig. 6.4; again a common feature is the large amount of primary production taken by the bacteria. The bacteria alone accounted for at least half of the carbon flow from the phyto-plankton, relatively most at station M and least at station E5. In this study,

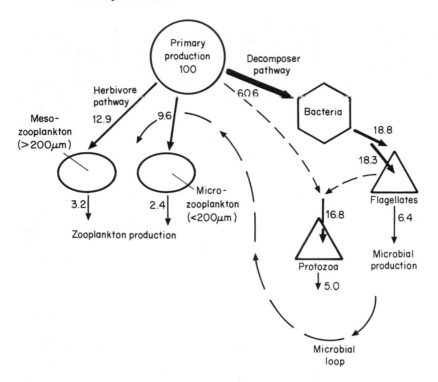

Fig. 6.4 A carbon budget for a mixed water English Channel station in summer during the decline of a phytoplankton bloom. All values are expressed as a percentage of the total carbon flow through consumer populations. Almost two thirds of the primary production enters the decomposer pathway. Of the 18.8% appearing as bacterial production, 18.3% is consumed by flagellates. Protozoan consumption is 16.8%. Production from the flagellate and protozoan components (microbial production) is twice that from the zooplankton components, an unquantified fraction of this microbial production may pass back to the herbivore pathway via the so-called 'microbial loop'. (From Newell and Linley (1984) and incorporating material from other sources.)

almost all of the bacterial production was found to be taken by small flagellates, suggesting that the bacterial standing stock might be controlled by flagellate grazing. This is supported by several other studies (e.g. Azam *et al.*, 1983). As the total heterotrophic demand estimated by Newell and Linley (1984) was considerably greater than the likely figure for primary production, they suggested that the heterotrophs in August were being sustained partly by the stock of organic carbon built up earlier in the year. It should not be assumed that bacterial activity always accounts for such a large fraction of primary production as was found in the English Channel in August. It would seem that this occurs under conditions of pulses of production, such as the temperate spring bloom. For example, Lancelot and Billen (1984) found that in the southern North Sea, there was a rapid response by bacteria to the presence of algal extracellular products. Bacterial numbers rapidly increased in response to an increase of the extracellular pool in the spring bloom, and it was estimated that 44–68% of the total primary production in the months of April, May and

June was used by the bacterioplankton.

It may be supposed that much of the production by bacteria, flagellates and large protozoa shown in Fig. 6.4 finds its way to the larger zooplankton, and thus back on the pathway leading ultimately to fish. Azam *et al.* (1983) referred to what they call the microbial loop, which is responsible for taking a large fraction of the primary production and respiring much of it, but then passing some back to the larger forms as production. This involves both the grazing of flagellates on bacteria and grazing by the larvacean *Oikopleura* on bacteria (King *et al.*, 1980). At station M, about 11% of the original primary production might pass back to the larger zooplankton from the flagellates and protozoa (Fig. 6.4).

In a different sort of system, Sorokin and Mikheev (1979) have found that bacteria are also very important in the Peru current upwelling. They investigated a patch of recently upwelled water separated by a front from the general water mass, said to be produced by 'quasi-permanent upwelling'. Between 20 and 50% of the primary production was not grazed by zooplankton (or by fish), but instead went either into solution as dissolved organic matter or sedimented out of the water column as detritus. At some stations there was free hydrogen sulphide in the water below 40 m. The anchoveta (*Engraulis ringens*) was identified as a significant grazer on the phytoplankton. Walsh (1981) also described the carbon flow in the Peru current upwelling, and his paper is particularly interesting as it provides a comparison of the system both before and after the collapse of the anchoveta fishery, which occurred in the early 1970s. A simplified version of the food webs described by Walsh is shown in Fig. 6.5. Before 1970, the web was simpler than after the collapse of the fishery, and with a substantial yield to man via the anchoveta. The bacterioplankton was estimated to take some 20% of primary production, via the detrital pool. After the fishery collapsed, euphausiids and sardines became relatively important, but the most striking change in the flow of carbon was the increase in phytoplankton production which went directly to the detrital pool, with an even greater relative increase in the carbon lost from the food web to the sediments. The yield to man was now a fraction of the earlier yield. Surprisingly, the bacterioplankton was now estimated to take only 2% of the initial primary production. Walsh does not indicate any output from the bacterioplankton such as has been suggested by Azam *et al.* (1983).

Good quantitative data seem to be non-existent for truly oceanic water columns. Roger and Grandperrin (1976) gave the following qualitative description of the food chain in the tropical Pacific:

phytoplankton
$\downarrow$ euphausiids → micronektonic fish → tuna
small zooplankton

Even in the absence of quantitative information, it is clear that the commercial yield (of tuna in this case) is likely to be much smaller than in an area where the harvested fish feed at least in part directly on phytoplankton (e.g. anchoveta). This follows from a consideration of the food chain length alone, but the relatively low annual primary production in the open ocean would reinforce this.

Everson (1984) has summarised the major food chain relationships of

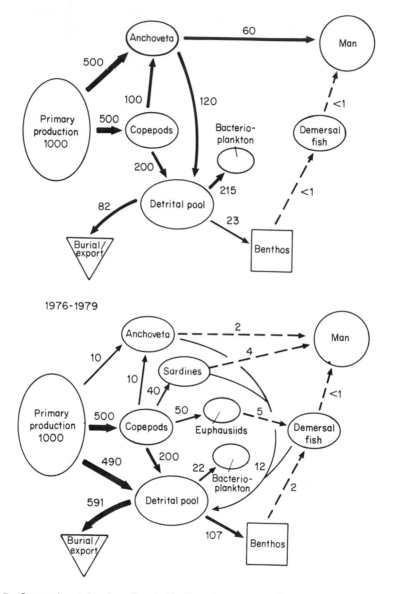

Fig. 6.5 Comparison of carbon flow in the Peru Current upwelling ecosystem before and after the collapse of the anchoveta fishery. (Redrawn and simplified from Walsh, 1981.)

Antarctic waters, as shown in Fig. 6.6. Again, the information is qualitative, but what is striking is the shortness of the food chain from the phytoplankton to animals used by man. Clarke (1985) commented that the very simple, linear picture such as indicated in Fig. 6.6 is unrealistic in that it omits many zooplankton consumers and the microbial loop. This is quite true, but it remains

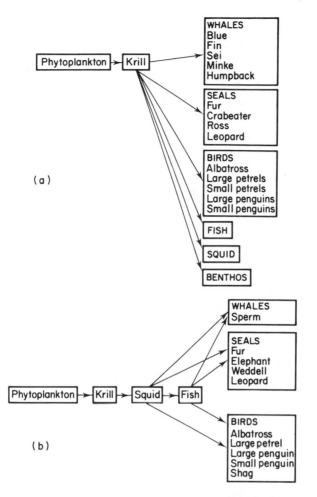

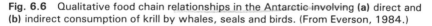

Fig. 6.6 Qualitative food chain relationships in the Antarctic involving **(a)** direct and **(b)** indirect consumption of krill by whales, seals and birds. (From Everson, 1984.)

impressive that in the Southern Ocean there can be such a short chain from the very small to the very large organisms. What is needed now is quantification of the relative importance of the various links in the food web.

Food chains and fisheries

The basic pattern

Ryther (1969) summarised information on marine production, dividing the ocean into three provinces: the open ocean (oligotrophic), and the coastal zone and upwelling areas, which were more eutrophic. His conclusions concerning productivity, food chain length, ecological efficiency and fish production are shown in Table 6.1. The factor leading to a shorter food chain in more eutrophic

Table 6.1 Productivity and food chain characteristics of the oceanic provinces (Ryther, 1969).

Province	Per cent of ocean area	Productivity (g C m^{-2} a^{-1})	Productivity (10^6 t C a^{-1})	Trophic levels	Ecological efficiency	Fish production (t fresh weight a^{-1})
Open ocean	90	50	10.3	5	10%	16 × 10^5
Coastal zone	9.9	100	3.6	3	15%	12 × 10^7
Upwelling areas	0.1	300	0.1	1.5	20%	12 × 10^7

waters was identified as the larger size of organisms involved, beginning with relatively large diatoms instead of small flagellates. In this context, the food chain length was specifically defined as the chain leading to commercially useful animals. By pointing to the short food chains and supposedly greater ecological efficiency in coastal and in upwelling areas, Ryther was able to suggest a mechanism which might allow the support of larger fish yields based on smaller amounts of primary production. The ecological efficiency was considered as being likely to be higher in eutrophic waters, although there was no concrete evidence for this, and as we shall see this idea has been challenged.

The fish production figure given in Table 6.1 is not an independent estimate, but is a calculation based on the figure for primary production and food chain length and efficiency. Ryther points out that this is not the same as potential yield: the fish will still have to be harvested. Also, his summary is vulnerable to the problems which beset all such attempts at general statements, in that the figures rely on data which themselves are often rather inadequate first approximations. However there is no doubt that the nature and dynamics of the food web are of great importance in governing eventual potential yield to man, and this will now be explored further.

Food web structure

It will by now be clear that under what may be called quasi-steady state conditions, where environmental conditions are fairly constant, the primary producers in the plankton tend to be very small. These conditions obtain widely in the open ocean at mid-latitudes and away from the boundary currents and upwellings. The autotrophic organisms favoured under these conditions are typically very small (< 10 μm), with the characteristics of high nutrient uptake capacity at low ambient concentrations, and the potential for very rapid growth. More turbulent conditions favour autotrophs which are somewhat larger. This is exemplified in the study by Holligan *et al.* (1984a,b) in the English Channel where stratified conditions favoured small flagellates, dinoflagellates were favoured in the frontal region and diatoms in tidally mixed waters. The dominant dinoflagellate was *Gyrodinium aureolum*, with a volume of 3847 μm³, while the dominant diatom was *Rhizosolenia stolterfothii* with a volume of 48 400 μm³.

The reasons for this rather general phenomenon are not fully understood, and it is most unlikely that any one factor is solely responsible for the size variations found in phytoplankton populations. As was pointed out in Chapter 3, the level of nutrient availability is widely thought to be important, with small and large cells being favoured in low- or high-nutrient conditions, respectively. This may be particularly true when light is considered as well – Parsons and Takahashi (1973) suggested that both nutrients and light had to be in good supply to favour large cells. Another factor thought to be involved is the velocity of vertical water movements; higher velocities are said to favour larger cells which have a greater tendency to sink. Since vertical water movement is associated with nutrient renewal in the euphotic zone, it is difficult to separate the two. Over continental shelves, turbulence may act to sustain diatom populations by causing re-seeding of the water column from cells or resting stages on the bottom. Wherever there is relatively strong vertical movement, larger phytoplankton seem to be more prominent, at least at some times.

The point which is of significance here is that under steady-state conditions the primary producers are small, while as environmental heterogeneity increases so larger cells become important. This heterogeneity may arise because of the 'pulsing' in time associated with seasonal changes at higher latitudes, or because of the spatial discontinuities caused by upwellings. Further, in upwellings there may be temporal heterogeneity as well.

The classical concept of food chains suggests that they are limited in length by the progressive dissipation of energy at successive trophic levels (the energy limitation hypothesis). Other possibilities are that since predators tend to be bigger than their prey there is an eventual limit imposed by size constraints, or that the advantages of feeding low in the food chain (because of the greater abundance of energy at lower trophic levels) acts to keep food chains short. Pimm (1982) analysed 26 food webs recorded in the literature, giving 56 food chains. He found that the modal length of the chains was of either 3 or 4 trophic levels, accounting for 41% and 43% of cases respectively. He showed that the number of trophic levels was fewer than would be the case if the food webs had been arrived at by random processes, and concluded that food webs did have 'structure' exemplified by a tendency to shortness of chain length.

Pimm considered that the available explanations of short food chains were inadequate. In particular, the energy limitation hypothesis would lead to the prediction that food chains should be longer in more productive situations, and there is no evidence of this. Indeed, it may be noted that for planktonic systems, the reverse would appear to be true. The steady state of the open oceans at mid latitudes is associated with reduced primary production and longer food chains, while the non-steady state of coastal waters, upwellings and especially the Antarctic is associated with enhanced primary production and shorter food chains.

Pimm and Lawton (1977) produced an alternative explanation for the restricted length of food chains. They investigated model systems using Lotka–Volterra equations to construct food chains of varying length, and judged the fragility of these by estimating how long the populations took to return to equilibrium after disturbance. This 'return time' increased with the number of trophic levels, and it was suggested that this might be the underlying cause of short food chains. Ultimately, a food chain must be limited by energy input, but perhaps this 'ultimate' limit is never reached simply because systems which approach it are too fragile to persist in a real world of disturbance. The clear corollary of this is that as frequency of disturbance increases so should food chains be shorter, and this is just what seems to be the case in planktonic food webs.

There is another consideration arising from food web models which seems to be relevant. DeAngelis (1980) showed that the resilience (i.e., the ability of a system to recover from perturbation) is inversely related to the time that a given unit, whether of energy or matter, spends in the system. In more everyday terms, this means that a high productivity/biomass (P/B) ratio among the components of a food web will enhance resilience because of a higher through-put of energy. These circumstances will be met among smaller rather than larger organisms, because of the higher rates of activity per unit of protoplasm in smaller organisms. Combining these observations with those of Pimm and Lawton (1977), webs with smaller organisms might allow longer food chains before their fragility becomes excessive.

The following speculative synthesis may be advanced. The environment of coastal waters (and especially those of temperate and high latitudes) and of upwellings favours large cells in the primary producer trophic level. Here not only is there an abundance of nanoplankton, but frequently an abundance of netplankton as well. At the same time, environmental unpredictability renders long food chains fragile, and they therefore tend to be short. Short food chains based (at least in part) on large producers are promising for fisheries. In contrast, the near steady state environment of the open ocean gyres favours small cells among the primary producers. This in turn may allow for the longer food chains which are observed here, and this means that of the already smaller energy input a relatively reduced fraction may be available as commercial fish.

As was appreciated by Ryther (1969), the nature of pelagic food chains is important in determining the yield to man from a fishery. The recent work of Pimm (1982) and others has confirmed the importance of food chain structure in this respect. Productivity as such has often been emphasised, but it is incre-asingly apparent that productivity is only part of the story, and perhaps not the most important part.

Food chain efficiency

It has been thought that the food chain efficiency of webs in more productive regions would be higher than that of webs in less productive regions (Ryther, 1969). Cushing (1973b) has estimated what he calls a transfer coefficient, the ratio of production in two successive trophic levels. This will be related to the ecological efficiency, though it may not be the same. Cushing used estimates of primary and secondary production in the Indian Ocean, in a variety of circum-stances from vigorous upwelling (high production) to the stable open ocean (low production). The mean value was very close to 10%, but the coefficient ranged from > 30% in unproductive to < 5% in productive waters (Fig. 6.7). Black-burn (1973) reported evidence of a similar inverse relationship between transfer or ecological efficiency for the tropical Pacific ocean. It seems therefore as if the relative amounts of energy passing up food chains may vary in a systematic way, being high in steady state conditions and low in fluctuating ones.

This is probably caused by the effects of lags in population growth. Individual organisms will take a finite time to respond to resource increase by growth and reproduction, and the larger they are the longer this lag is likely to be (Peters,

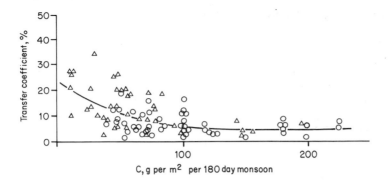

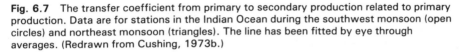

Fig. 6.7 The transfer coefficient from primary to secondary production related to primary production. Data are for stations in the Indian Ocean during the southwest monsoon (open circles) and northeast monsoon (triangles). The line has been fitted by eye through averages. (Redrawn from Cushing, 1973b.)

1983). In a food chain where producers and consumers are much the same size, with similar growth rates, consumer populations will track those of producers relatively closely, leading to what may be called a high degree of coupling in the food chain. This underlies the relatively low variability of nanoplankton biomass noted by Malone (1980). In contrast, where producers are diatoms with generation times of hours or days and consumers are copepods with generation times of weeks or months, there is scope for considerable food chain decoupling. Under these circumstances the planktonic food web may export substantial quantities of organic matter to the benthos, where it may be incorporated into organisms useful to man.

7

The biology of commercially important fish

In previous chapters, an account was given of the physical background to plank-tonic production, and of the organisms and processes involved in that produc-tion. How some of the energy fixed by phytoplankton may enter a food web, and the properties of food webs as they affect fish have been considered. It remains in this and following chapters to describe selected features of the biology of some marine vertebrates which are important in governing their response to harvesting by man. This will of necessity be a highly selective account, and more comprehensive treatments will be found in the works cited in the references. The approach adopted here is to discuss food and reproduction with reference to selected examples of fish, drawing on knowledge of the cod (*Gadus morhua*), plaice (*Pleuronectes platessa*), clupeoids (e.g. herring, *Clupea harengus*, and anchovy, *Engraulis* spp.) and tuna (e.g. *Thunnus* spp.). The first section of this chapter deals with the term 'stock' as it applies to fisheries.

Species and stocks

Biologists classify living organisms into taxonomic groups at various levels, the most frequently used being that of species. A full discussion of what a species may be is beyond us here, but a useful notion is that all the members of a species should be at least potentially capable of gene exchange. Those members which actually constitute a breeding group with unimpeded gene exchange may be said to belong to a population; commonly there are small genetic differences between populations which are insufficient to constitute a reproductive barrier. These differences may be adaptive in that they suit members of the population to local conditions; if this is so then they will tend to be maintained even in the event of some genetic mixing from other populations. Fisheries biologists recog-nise groups within species which they call stocks, and in fact this term is synony-mous with population as just defined. That is, the species are divided into local populations which show genetic differences between them, and these diff-erences may be adaptive.

The identity of fish stocks in the sea is maintained principally by separation of the species into populations which gather in different places for spawning. The eggs and larvae are therefore segregated in space (and sometimes in time as well) with their own pattern of drift in the plankton, nursery grounds, and migration to the adult stock as they grow. Adult stocks may mix on the feeding grounds, but separate out again to spawn.

Another term which merits brief discussion is recruitment. To a fisheries biologist, recruitment occurs when a young fish joins the adult, fished stock.

This may happen some considerable time after spawning, thus Cushing (1967) writes of the Downs herring in the North Sea recruiting at an age of 3 to 4 years.

Food

Bottom feeders

Plaice and cod feed, as adults, on or near the sea bottom. Plaice feed on benthic molluscs and crustaceans, they have crushing teeth in the throat (the pharyngeal teeth) with chisel shaped teeth on the jaws. They also feed on polychaetes, particularly as newly settled juveniles. In some areas they may rely on such food as the siphons of infaunal bivalves, these represent a rapidly renewable resource as the prey animal survives and regenerates new siphons. Cod feed on other fish and on crustaceans, and have strong, pointed teeth for grasping their prey. Both of these examples are at some distance up the food chain from the primary producers.

Pelagic feeders

The herring is representative of a large and important group of fish (clupeoids) which feed on planktonic organisms in the water column. The food of *Clupea*

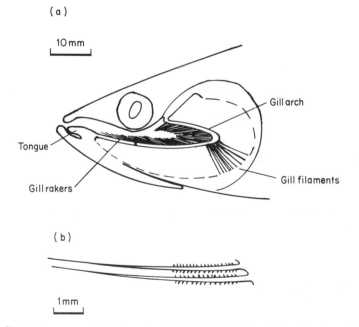

Fig. 7.1 *Clupea harengus.* (a) the head with the left gill cover removed to show the first of four gill arches. Each bears gill filaments pointing outwards (only some shown) and gill rakers pointing inwards. Each raker (b) is a flattened elongate projection, most bear tiny spines from the base for a varying distance along the length of the raker. The spines are borne on the inner edge and are found along the whole length of the rakers in the angle of the gill arch, reducing in number until the rakers at the ends of the arch lack spines altogether.

harengus itself was discussed in Chapter 6. The animal is not an indiscriminate feeder, but selects food visually. It stops feeding in complete darkness, although moonlight is sufficient for feeding to occur (Parrish and Saville, 1965). The prey is engulfed in its surrounding water, prey particles are retained on the gill rakers. These are stiff projections from the inner side of the gill arch (Fig. 7.1), of which there is a double row on each arch. They interlock to form a sieve. In many clupeoids feeding is more a process of filtering, and will proceed in the dark (Blaxter and Hunter, 1982). The clupeoids tend to be near the base of the food chain, especially so in the case of phytoplankton feeders such as anchovies.

Tunas locate food largely by sight, and are said to be restricted in their distribution to water clear enough for visual feeding but with enough phytoplankton to sustain their food (Sund *et al.*, 1981). They are widely distributed through the tropical and subtropical oceans, often particularly associated with upwellings. They are active predators at (or near) the top of the natural food chain, taking as food items fish, molluscs and crustaceans in the size range 10 to 100 mm. Collectively these may be thought of as being planktonic or micronektonic. As tuna are principally visual feeders, they take few of the potential food items present near the surface at night as a result of vertical migration. In the area of the Baja California upwelling, the yellowfin and skipjack tuna feed largely on larvae of the red crab, *Pleuroncodes planipes*. They are restricted in this region by low temperatures: if the surface temperature is below 20°C the tuna do not penetrate the water mass, and exploitation of this food resource is limited until such time as temperatures rise. When feeding on red crab larvae, tuna are unusually near the base of the food chain.

Larval food

As larvae, all these fish are planktonic. Larval clupeoids feed on phyto- and zooplankton, depending on availability. Copepods are evidently especially important for some. Anchovies will feed successfully on dinoflagellates (Blaxter and Hunter, 1982), and indeed these motile phytoplankters seem to be the best food for early larvae of *Engraulis ringens* (Walsh, *et al.*, 1980). This is presumably because the larvae are visual predators, and may require motile prey to trigger feeding.

The nature of larval food and its importance is well known for plaice. The newly hatched larvae are sustained by remaining yolk, but while this is still being used they begin feeding on diatoms. At the end of the yolk sac phase, there is a fairly abrupt transition to taking zooplankton, with larvaceans (such as *Oikopleura*) being particularly important (Shelbourne, 1953, 1957). Larvaceans are soft-bodied, slow-moving and yet conspicuous by virtue of the lashing tail, and are supposed to represent particularly favourable prey. The transition phase is considered to be particularly critical, if adequate food is not available, growth becomes slow and the larvae may starve.

Reproduction and recruitment

The timing of spawning

Bony fish generally use external fertilisation, shedding the gametes into the sea in the cases considered here. Eggs may develop on the bottom (herring) or in the water column (anchovy, cod, plaice, tuna), hatching ultimately into planktonic larvae. The aspect of reproduction which concerns us here is its timing in relation to environmental phenomena.

The timing of spawning appears to be adjusted so that the larvae which eventually result are in a favourable environment, especially with respect to food. In tropical clupeoids, the small species associated with upwellings link their spawning to times of upwelling, which provides food for the spawning adults and the subsequent larvae. Temperature is involved in controlling time of spawning, in some El Niño periods spawning by the anchoveta *Engraulis ringens* may be advanced in the warmer water (Walsh *et al.*, 1980). However the picture is complicated, for in another El Niño episode spawning by this fish was very slight and gonads were eventually resorbed. It may be that adequate food is necessary for initiating spawning of previously matured gonads, this could offer a means of directly linking spawning to the availability of food (Blaxter and Hunter, 1982). Thus, in fish such as these, we can discriminate between environmental variables which may set in train the events leading to future spawning, and those which may act more immediately on the mature fish to induce or inhibit spawning. Between years, there may be substantial variation in the time of peak spawning, over as much as several months.

For fish such as the herring or plaice, there is no opportunity to modulate spawning according the larval food supply, simply because this cannot be assessed by the spawning fish. In herring, the spawning season is about two to three weeks, while in plaice and cod it is of up to three months duration (Cushing, 1975). From surveys of eggs in the plankton the time of peak spawning can be estimated; in plaice of the southern North Sea it is 19 January with a standard deviation of 7 days. While the spawning season for the population as a whole is months long, the spawning of an individual fish will be shorter. Plaice are said to spawn the contents of the ovary in two to four weeks (Jenkins, 1925). Information on the time of spawning in herring comes from the timing of the appearance of newly hatched larvae in the plankton. Evidence is also obtained from the state of caught herring, whether ripe or spent, and from catches of fish such as haddock which have been feeding on the spawn.

Figure 7.2 shows the seasons of spawning of herring around the British Isles, and it can be seen that there is a progression from spawning in late summer/autumn in northern waters to winter in southern waters. Cushing (1967) related this to the time when hatching larvae will need food. Hatching takes place in about 2–3 weeks depending on temperature, so that winter spawners are dependent on the phytoplankton of the spring bloom, commencing around February in the English Channel and southern North Sea. Autumn spawners depend on the autumn production peak in the North Sea (see Chapter 4). Similarly, plaice in the southern North Sea spawn so as to enable the larvae to use the plankton bloom occurring later in the year.

Cushing (1975) suggested a generalisation about the timing of spawning. In tropical and subtropical waters the time of spawning is not precisely fixed; this is linked to the nature of production, which is either quasi steady-state or capriciously variable. Thus, tuna larvae can be found through the tropical Pacific for much of the year, while the clupeoids of upwellings effect some regulation of spawning depending on immediate food conditions. In mid- and high-latitudes, with a comparatively regular seasonal pulse of production, spawning occurs at

Fig. 7.2 Spawning times of herring stocks in the Irish Sea and North Sea. (From information in Cushing, 1967.)

such a time as to maximise the chances of the larvae finding sufficient food in a subsequent plankton bloom.

The 'match-mismatch' hypothesis

If long-lived fish such as cod, plaice and herring are aged they can be assigned to year classes, and these are found to vary substantially. The relative success of recruitment in a year is reflected in the wide variations of year classes. The ratios of best:worst years have been calculated for a variety of fish, in clupeoids this ranges between 2:1 and 25:1 and in flatfish between 2:1 and 16:1 (Roff, 1981). There may be substantial variation within a species, depending on the population (or stock). It is widely considered that such variations are due to environmental factors acting on the larvae, and many correlations have been made between year-class-strength and climatic variables (Cushing, 1982). The mediating factor in this is likely to be the availability and quality of food. In waters round Britain, the timing of the spring bloom may vary by up to 6 weeks (Cushing, 1982), and it may be imagined that there is considerable scope for getting it wrong in the process of spawning eggs in advance of such a plankton bloom. A large element of reproductive success will then be the degree of match of the timing of larval need with availability of food. This has been expressed graphically by Cushing (1982), the bell-shaped curves in Fig. 7.3 represent the

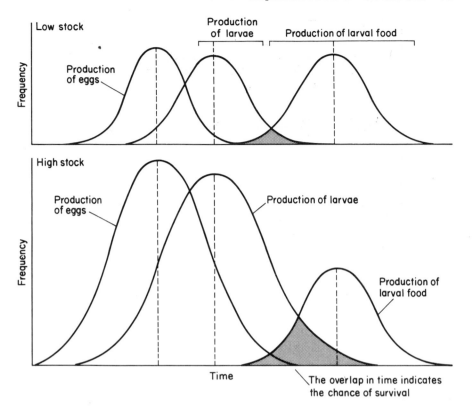

Fig. 7.3 Diagrammatic representation of the 'match-mismatch' hypothesis. The timing of the third peak is supposed to be most variable, leading to the possibility of better or worse match between larval demand for food and its availability as the curve of food production shifts back and forth along the time axis from year to year. In any one year, low stock (upper graphs) will accentuate the effect of mismatch. (From Cushing D.H. (1976). The impact of climatic change on fish stocks in the North Atlantic, *Geographical Journal*, **142**, 216–27.)

time sequences of egg production, larval numbers and food availability. The peaks of the first two distributions are relatively constant, while that of the third is subject to larger variations. Another feature of this is that as the size of the parent stock decreases so the spread of larval abundance is likely to decrease, making the match between larvae and food worse.

There is a long series of records available for plaice in the North Sea. Cushing (1982) referred to three exceptionally cold winters since the commencement of records: those of 1928–9, 1946–7 and 1962–3. Two of these were followed by very high recruitment of plaice, that of 1963 being the highest on record. It is not clear why 1946–7 should not have produced unusually good recruitment, but for the other two years Cushing suggested that because development was delayed by unusually low temperatures an unusual degree of match was obtained between larval abundance and food availability. This raises the unanswered question of why plaice should apparently usually breed so as to produce a less than optimal match with food availability for the larvae.

Density-dependent effects

The factors causing changes in recruitment discussed above may operate in a largely density-independent manner. That is, the severity of the effect is unmodified by the density of the population on which it acts. There is, however, evidence of density-dependent effects on recruitment in a wide range of fish, with the magnitude of the density-dependence varying between species and stocks. The evidence for this, and possible mechanisms, have been reviewed by Cushing and Harris (1973) and Shepherd and Cushing (1980).

The eggs and larvae of plaice and cod, and the larvae of herring, drift in the plankton. Spawning grounds are well-defined spatially, and are positioned such that the drift of eggs and larvae is advantageous. In the example of the southern North Sea plaice discussed by Cushing (1975), eggs spawned in the region between the estuaries of the Thames and Rhine drift north eastwards. The metamorphosing larvae settle along the Dutch coast and enter the Waddensee, a large estuarine region. It has been suggested that the larval phase represents a period during which there may be density-dependent mortality.

Some of the evidence for a density-dependent response comes from the relationship of recruitment to parent stock, the stock-recruitment relation. To obtain this information, data on the sizes of stock and the subsequent recruitment are needed for a number of years. Expressed graphically, these often resemble the points in Fig. 7.4a, for the Arcto-Norwegian cod (Garrod and Jones, 1974). The curve in Fig. 7.4a is the best fit mathematically, the pecked lines represent the 95% confidence limits for the curve. Figure 7.4b shows a series of such curves for different fish, and illustrates the point that the stock-recruitment relation varies between them. It is true also that this relation varies

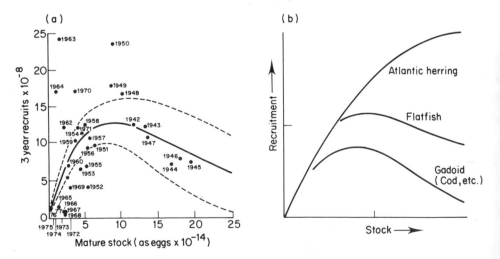

Fig. 7.4 **(a)** Estimates of recruitment plotted against stock (as numbers of eggs) in the Arcto-Norwegian cod population. The solid line is a least-squares regression curve constrained to pass through the origin of the graph, the broken lines are the 95% confidence limits of this curve. (From Garrod and Jones, 1974.) **(b)** General stock-recruitment curves for three examples of groups of fish populations, the axes are arbitrary. (From Cushing and Harris, 1973.)

between stocks within a species. The degree of density-dependence thus seems to vary quite markedly between different fish.

Shepherd and Cushing (1980) have pointed out that growing fish larvae may be said to pass through a series of 'predatory fields', in each of which they are vulnerable to predators taking prey in a particular size range. The smaller they are, the more abundant will be their potential predators, because of the tendency for smaller animals to be progressively more abundant (Peters, 1983). In particular, after metamorphosis the nature of predation on the young fish changes. Now even at constant rates of mortality, the more quickly a cohort of larvae grows through a particular predatory field, the more individuals will emerge safely. Conversely, if they are delayed in a predatory field fewer individuals will survive. Such a delay may occur if growth is depressed through shortage of food, and this offers a possible mechanism for density-dependence. In years when larvae are superabundant, shortage of food may lead to high mortality. Obviously this is a very simple statement of a complex problem, for example food abundance and quality is also variable, but this does not remove the likelihood or importance of density-dependent factors.

Environmental changes and fish populations

The Peru current upwelling

From the above discussion, it will be clear that fish populations rely on a more or less capricious environment. The most spectacular example of this is the Peruvian anchoveta, *Engraulis ringens*. This fish is dependent on the production

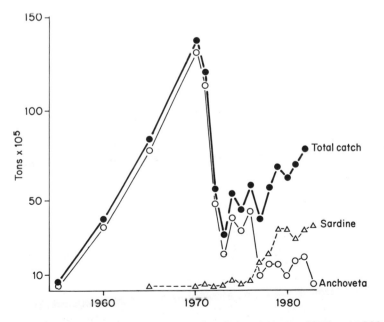

Fig. 7.5 Fish catches in the Peru Current upwelling fishery between 1955 and 1983. The points for 1983 are estimates. (Drawn from data in Gulland and Garcia, 1984.)

associated with the upwelling of the Peru current. During episodes when the upwelling ceases (El Niño) their spawning may be disrupted, and the adult stock may be dispersed southward. In addition to this, the stock has been subjected to heavy fishing pressure. Figure 7.5 shows the annual catches of major species in the fishery for various years between 1955 and 1983. The anchoveta fishery was at its height in 1970–71, after which it collapsed and has not recovered. At the same time there has been an increase in the catch of sardine, but this does not compensate for the loss of the anchoveta (Gulland and Garcia, 1984). The collapse of the fishery for anchoveta seems to have been due both to fishing pressure and the effects of successive El Niño episodes. It seems to be impossible to disentangle the major contributory causes and say which are the more important factors. Similar changes have occurred in other upwelling fisheries, and certainly off California there have been changes in the balance between anchovy and sardine populations quite independent of fishing (Gulland and Garcia, 1984). However there can be little doubt that the fishing pressure exerted by man has played an important part in the collapse of the Peruvian anchoveta fishery (Walsh, 1981; Cushing, 1982).

The English Channel

The changes which have occurred in the western English Channel in this century have been particularly well described. As a result of long term observations from Plymouth, reviewed by Southward (1980), we know that between about 1925 and 1935 there was a series of changes. Average sea temperature increased by about 0.5°C, the winter maximum of dissolved phosphate decreased, and there were changes in the zooplankton. There were also changes in the demersal fish populations towards warmer water species, and a disappearance of herring, apparently replaced by pilchard. Then in the late 1960s the changes went into reverse, except that the pilchard were replaced as the dominant pelagic fish by mackerel, and not by herring. Except for this difference, the system now closely resembles that of the 1920s. A striking fact is that there was no major change in the levels of primary production associated with the shift of the system from one state to the other.

There is a variety of possible causes for these major shifts, including nutrient control, competition between different fish species, and climatic change. These of course are not mutually exclusive, and some combination of all three might be involved. Southward (1980) considered that both direct and indirect climatic effects are responsible, producing shifts in species distribution and changes in water circulation patterns, respectively. There is evidence from the history of the Cornish pilchard fishery of changes in pilchard abundance over the past 400 years (Southward, 1974). This suggests that similar changes to those well documented in the 20th century are the norm for the western English Channel, and that as far as commercial fish are concerned, equilibrium is precarious.

The North Sea

The herring stocks in the North Sea collapsed in the late 1960s, at least partly as a result of increased fishing pressure (Murphy, 1977; Cushing, 1980). In 1962 there began an increase in gadoids (the cod family), known as 'the gadoid outburst' (Cushing, 1980). Gadoids commenced a succession of ever better recruitments over twelve years, especially marked in cod. This might have been due to release of food no longer taken by herring, or to a relaxation of predation by herring taking the smallest cod. Either of these mechanisms would be encouraging for those who try to manage fisheries, in that the collapse of one fishery could be seen as releasing resources to another potential fishery. However the increase in gadoids may also have been due to climatic changes, and Cushing (1980) favoured this last explanation as being the most likely. If this is so, then it is not quite so comforting for fishery managers, for it suggests that the change from herring to gadoids was more likely to be fortuitous than as a result of release of resources.

Reproduction in an uncertain environment

In common with a great number of marine animals, many fish use the plankton for their eggs and larvae. The advantages of this may be to do with the wide dispersal obtained, although there are lots of examples of animals where these are foregone, and it has been argued that in invertebrates the degree of dispersal is in excess of what would be optimal. The topic has been reviewed many times, most recently by Grahame and Branch (1985).

Murphy (1968, 1977) suggested that the life span of clupeoids might be adapted to the degree of uncertainty of the environment. He claimed that there was a good correlation between length of reproductive span and the variation in spawning success, estimated by the ratio of highest:lowest spawnings. This ratio was 2 in the Peruvian anchoveta, with a reproductive span of 2 years, ranging to 25 in the Atlanto-Scandian herring with a reproductive span of 18 years. Roff (1981) re-examined the data used by Murphy and reached rather different conclusions. He pointed out that the anchoveta had a much greater variation in spawning success than Murphy had taken into account, due to events following a severe El Niño subsequent to Murphy's first paper. Moreover much of the rest of the correlation depended on one species, *Clupea harengus*. Using data for flatfish as well as clupeoids, Roff (1981) came to the conclusion that variation in length of reproductive span was more likely to be a correlate of the age at first maturity. Thus fish with early maturity tended to have a short reproductive life, while those with late maturity tended to have a long reproductive life. For a species to persist, sufficient individuals must replace themselves in the population, and it may be argued that if maturity is delayed, a necessary concomitant of this is longer reproductive life, or higher fecundity. Roff (1981) suggested that within groups of fish such as the clupeoids and flatfish, it is the reproductive span which varies in this way.

Fecundity (that is, the number of eggs spawned, either per occasion or per lifetime) also varies between groups of fish. Cushing (1973a) found evidence for a relationship between fecundity and the degree of density-dependence in the stock-recruitment relation. There seems to be a progression from fish with low fecundity and low density-dependence, such as clupeoids, to those with high fecundity and high density-dependence, such as gadoids. This has clear implications for the management of fisheries, with clupeoids being relatively susceptible to fishing pressure, and gadoids resilient.

8
Whales Cetacea

The mammals are primarily terrestrial animals, with warm blood, fur, and live-born young suckled on milk. Many show a degree of adaptation to aquatic life, including the platypus, which is an exception to the general rule in laying eggs. The order Cetacea, placed among the Eutheria or placental mammals, have become fully aquatic. The Cetacea comprise the whales, dolphins and porpoises, animals ranging in size from about 2 m to 30 m in length. They are mostly marine, but some small species live in fresh water. The animals which concern us here are some of the large marine forms, i.e. the baleen whales (Mysticeti) and one of the toothed whales (Odontoceti), the sperm whale. It is only in the Mysticeti that there are animals which may really be said to be plankton feeders. The choice has been made simply because the baleen and sperm whales have formed the basis of large commercial fisheries. The factors which have made them attractive to man are the valuable products to be obtained, and their highly aggregated distribution, which facilitates hunting.

Nomenclature

Whales are frequently referred to by their 'common' names. Therefore it seems useful to give a list of common and Latin names together with some information on distribution for the whales mentioned in this chapter. Some species have synonyms, the names given here follow Gaskin (1982).

Suborder Mysticeti

Family Eschrichtidae (grey whales)

Grey	*Eschrichtius robustus* (12 m, 13 m)	North Pacific

Family Balaenidae (right whales)

Greenland Right, Bowhead	*Balaena mysticetus* (15 m)	Arctic
Black or Biscayan Right	*Eubalaena glacialis* (15 m)	Widespread

Family Balaenopteridae (rorquals)

Humpback	*Megaptera novaeangliae* (15 m)	Widespread
Blue	*Balaenoptera musculus* (25 m, 26 m)	Widespread

Fin	*B. physalus* (21 m, 22 m)	Widespread
Sei	*B. borealis* (15 m, 16 m)	Widespread
Minke	*B. acutorostrata* (8 m)	Widespread

Suborder Odontoceti

Family Physeteridae

Sperm	*Physeter catodon* (15 m, 11 m)	Widespread

Family Delphinidae

Pilot whale	*Globicephala melaena* (6 m, 5 m)	Widespread, not North Pacific

Figures in brackets are the average lengths of the animals, for males and females respectively if the sexes are dimorphic. In baleen whales the female is the larger sex while in sperm and pilot whales the male is larger.

Biology of whales

Migration

With the exception of the bowhead, baleen whales perform large seasonal migrations from high to low latitudes and back again. These allow the animals to feed in the rich plankton of high latitudes during summer, and to give birth (calve) in warmer water nearer the equator during winter. At least, this is the conventional wisdom, and certainly the advantages of feeding on the high-latitude plankton blooms are fairly clear. Blue, fin, humpback and sei whales taken at subtropical whaling stations in winter have little or no food in the stomach, and it seems that the period spent in warmer water is a non-feeding one. As the calves of the large whales are relatively small, they have a much higher surface area:volume ratio than do the adults, and with little or no blubber at birth it is supposed that they might suffer from hypothermia if born at high latitudes. However Gaskin (1982) questioned the assumption of a need to calve in warm water, pointing out that adults of small species (which would have similar surface area:volume ratios, and similar problems of heat loss) may penetrate very cold water. Some large species do indeed calve in cold water at high latitudes (e.g. *Balaena mysticetus*). The reason for migration may be to spend a time of food shortage in a metabolically less demanding environment, as discussed below.

Whatever the reasons for these migrations, they result in a considerable degree of separation of the populations. Southern hemisphere populations are near the equator when their northern counterparts are in high latitudes, and *vice versa*. Apparently crossing from one hemisphere to the other is rare, and there may be virtually complete reproductive isolation between the populations of what are regarded as a series of species. Mating takes place after the calving season.

Sperm whales are widely distributed, but do not penetrate the high Arctic. Males are larger than females (at maturity, three times the weight of a female of the same age) and may occur in mixed schools with females, or in 'bachelor schools'. Large bachelors are found at the highest latitudes, mixed schools and smaller bachelors at mid and low latitudes. It appears that only males take part in the full high-low latitude migration, and perhaps not all of them do so. Breeding schools are found at low latitudes.

Food and feeding

Sperm whales feed on squid and fish caught at some depth. Off Iceland they are known to take benthic fish from 500 m, in oceanic waters they turn more to squid. The size range of prey taken is considerable, from fish of 40 mm to giant squid several metres long. How the prey are caught is not known, an intriguing suggestion is that they are literally stunned by intense, focused sound waves produced by the whale (Norris and Mohl, 1983). This would certainly help to account for the fact that sperm whales seem to be able to feed when hampered by a very deformed lower jaw, further the prey in the stomach usually show no tooth marks.

The baleen whales are hugh filter feeders. Baleen consists of long, thin fringed plates of keratin, hanging like curtains from the upper jaw. Figure 8.1 shows the arrangement of the baleen plates and the tongue in the mouth of a baleen whale. Among rorquals (balaenopterids), the whale opens its mouth, engulfs a volume of water, and closes the mouth again. The floor of the mouth is depressed while the mouth is open, when the mouth closes it is elevated, forcing water out through the filter formed by the fringes of the baleen plates. Prey items are trapped in the filter and swallowed, the tongue being used to push food backwards in the buccal cavity. This method is known as 'gulping', whales may also perform 'skimming' when they swim slowly forward with slightly open mouth, a current of water passing into the mouth from the front and then out through the baleen. This is most often seen in right whales, and in the sei whale.

The right whales feed on zooplankton and micronekton, with copepods often featuring prominently in the diet. The pteropod *Limacina helicina* is said to

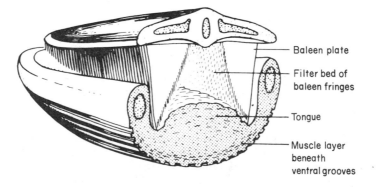

Baleen plate

Filter bed of baleen fringes

Tongue

Muscle layer beneath ventral grooves

Fig. 8.1 The sectioned head of a balaenopterid whale showing the organisation of the mouth. (From Bonner, 1980.)

form a major item in the diet of the Greenland right whale. Grey whales feed on invertebrates caught from the bottom, it is thought that they plough along the bottom stirring up sediments and feed on the disturbed animals by filtering them out.

The rorquals tend to differ in the food they take, depending upon their location. Northern populations take fish, squid and planktonic crustacea depending on the species of whale and the food available, but in the Southern Ocean the principal food is euphausiids. An exception to this pattern is the blue whale, which feeds on euphausiids throughout its range. In the Southern Ocean it feeds almost entirely on *Euphausia superba*. Another exception is the sei, which feeds on euphausiids at high latitudes in the Southern Ocean, and on amphipods and copepods at lower latitudes.

Size and metabolism

The large rorquals in the southern hemisphere, which migrate to the Southern Ocean, are noticeably larger than their supposed conspecifics in the northern hemisphere. In the case of the fin whale, the Antarctic form is longer by 8% and heavier by 30% than that of the North Pacific (Brodie, 1975). One view of the relatively large size of southern forms is that as this reduces the surface area of the body proportional to volume heat is conserved, and the massive size of southern blue and fin whales may then be seen as an adaptation to feeding in very cold water. However, these size differences are not apparent in the smaller rorquals, such as the minke. These animals spend longer in the Antarctic, with a longer feeding period. Brodie (1975) considered that the larger size of blue and fin whales in the southern hemisphere is an adaptation to a long period of fast: the large size implies a reduction of weight-specific metabolism, while conferring an ability to store large amounts of blubber. The rorquals, then, may be seen as adapted to harvesting the brief Antarctic plankton bloom, with the largest whale species as the best adapted. Migration to warmer water to calve is perhaps to be understood in terms of saving the adults' energy, as the cost of migration may be less than the cost of remaining in the cold Antarctic waters during winter when food is scarce.

Social organisation and population biology

The baleen whales are often in small schools, of less than 10 animals. Gaskin (1982) used the term 'school' to denote a group in which there is some social cohesion, while an 'aggregation' is a group of schools. He maintained that there are selective advantages in these animals operating in small loose groups, both in enabling them to feed with minimal mutual interference and in minimising the risk of attracting predators when on migration.

Odontocetes are more markedly gregarious, and sperm whales may be found in mixed sex schools of 20 to 40 animals, while males may occur as solitary individuals or in groups of up to 50. The breeding relationships of the mixed sex schools are not known, in particular there is controversy over whether sexual activity is confined to one large male (the 'school-master'), or to the young males. The schools of sperm whales are more structured than are those of baleen whales, and it has been suggested that the loss of the 'school-master' male to whalers may disrupt the breeding of a school. This must remain speculative, however, in view of our ignorance of the details of interactions in the schools.

The age of larger species of baleen whales may be determined by counting layers in the wax plug found in the outer ear canal. This is a far from perfect method, and for a time there was doubt over whether one or two layers were deposited each year - the initially accepted number was two. It is now accepted to be less than two each year and the method remains somewhat imprecise (see Gaskin, 1982, for discussion). It is particularly difficult in smaller species, including the sei and minke. Age in odontocetes is estimated by counting annual layers in the teeth.

The natural longevity of whales is considerable. Once past the juvenile stage, when predators such as sharks and killer whales may exert some mortality pressure, the larger whales seem to be relatively immune to predation. Data are scarce, but it is reasonably well established that fin and blue whales live for 40 to 50 years (Harrison, 1969).

Fecundity in Cetacea is low, as might be expected in animals which at their smallest are still relatively large for the mammals as a whole. The principal commercial species are towards the upper end of the size range, with the lowest fecundities and most protracted reproductive cycles. Once mature, females may calve every second or third year, or in the sperm whale every fourth year. Twinning is extremely unusual (< 1 % of births), and one or both twins are likely to die.

Harrison (1969) gave some comparative data for reproductive events in rorquals. Sexual maturity is said to be reached at about 5 years (blue, fin whales) or 2 years (minke, sei whales). Gestation lasts 11 to 12 months, lactation 5 to 7 months. Harrison pointed out that there has been confusion between the time of puberty (when there is no definite evidence that cetaceans may immediately become pregnant) and attainment of sexual maturity, when it is known that the gonads are active. Slijper (1962) gave similar figures for age at maturity in blue, fin, minke and sei whales, except that for North Pacific sei the age at maturity is 5 years. More recent investigations in rorquals are said to show that sexual maturity occurs later than was thought.

This disagreement is particularly clear in the case of the sei and minke whales. Horwood (1980) reviewed the population biology of the southern sei whale, giving female age at maturity as about 10 years. An interesting feature shown in this work is a decline in age at maturity, from 11 to 12 years in 1939/40 to about 9 years in 1968/9. Some of this may reflect changes due to whaling pressure. A similar change has been seen in the fin whale (from 6 to 10 years old at maturity) and minke whale (from 6 to 14 years), with some variation depending on the region of the Southern Ocean (Brown and Lockyer, 1984). At the same time, pregnancy rates have shown a tendency to increase. These changes are shown for some species in Fig. 8.2.

Gestation in sperm whales lasts 15 to 16 months, lactation 10 to 15 months depending upon the population (Harrison, 1969). Sexual maturity is reached at between 7 and 12 years in females, depending on location, and at 18 to 19 years in males (Brown and Lockyer, 1984). At maturity lengths are about 8 m (females) and 12 m (males). In the smaller odontocete *Globicephala melaena* (common pilot whale), females become sexually mature at 6 years (length,

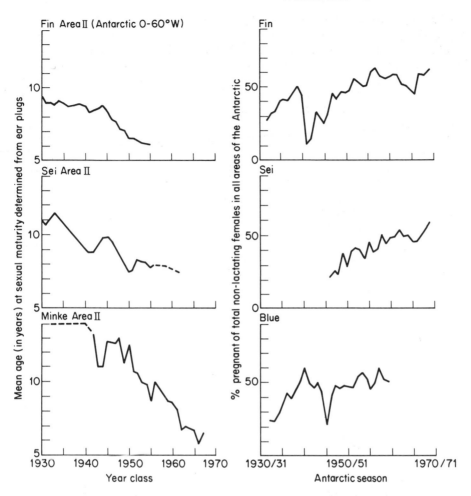

Fig. 8.2 Changes in age at sexual maturity and pregnancy frequency in some balaenopterid whales in the Antarctic. (From Brown and Lockyer, 1984.)

3.5 m) and males at 12 years (length about 5 m) (Harrison, 1969; Slijper, 1962). These times seem exceptionally long and Slijper commented that they must be regarded with some caution, although in the context of the mammals as a whole the larger whales are said to be precocious for their size.

Whales and the antarctic

A complete review of whaling is beyond the scope of the book; accounts may be found in Slijper (1962) and Gaskin (1982), to name but two. However a brief outline seems appropriate, before considering whales and the plankton-based food web of the Antarctic.

Whaling and its impact

Whaling is an old activity, going back to the stone age in Norway (Slijper, 1962). The principal prey was the right whale, and its hunting spread from Norway through maritime Europe, with the Basques of northern Spain taking whaling across to North America. As pressure increased, so right whales diminished in the northern hemisphere, and attention turned to sperm whales. This took whalers further afield, and southern right whales were exploited. All this activity was mounted from sailing ships from which small hunting boats were launched to harpoon whales by hand. In the mid-nineteenth century the harpoon gun was perfected, and this together with the use of steam ships allowed the full exploitation of southern whale populations, with the faster rorqual whales now being taken in great numbers. Before World War II the main prey was the blue whale, but when whaling recommenced after the war attention shifted successively to the fin, sei and minke as one after another became reduced in numbers. This is, of course, a progression down the size categories of rorqual species.

Uncertainty about whale population sizes, especially in the past, makes it difficult to quantify the effect of whaling. The relatively small-scale 'drives' against pilot whales, practised by fishing communities in Newfoundland and the Faeroes, have exerted a significant effect on Newfoundland populations of pilot whales (Gaskin, 1982). The efficient hunting of baleen whales has undoubtedly been devastating. This began in the Antarctic in 1904, by 1911 catches were already being affected by declining populations. The stern slipway factory ship (1924) allowed widespread oceanic exploitation in the Southern Ocean.

Right whale populations collapsed worldwide under the pressure of whaling, with some evidence of recovery now that pressure on them has been much reduced. The Greenland right is now represented by about 2500 animals, the black right by about 3500 (Gaskin, 1982). Among the rorquals, current global population estimates vary from about 3000 to 5000 (humpback) to several hundred thousand (minke).

Effects in the Antarctic

As mentioned above, several rorquals have shown a reduction in the age at maturity and an increased pregnancy rate (percentage of females pregnant) which has accompanied the overexploitation of the Southern Ocean stocks. This is shown graphically in Fig. 8.2. In the minke, changes began to occur before they were extensively hunted (Brown and Lockyer, 1984), the same is true of the sei (Horwood, 1980). It may be suggested that these changes represent a response by these smaller species to increased food availability as first one and then another baleen whale was overexploited, starting with the largest species.

Laws (1977) reviewed the changes in Southern Ocean whale stocks as a result of whaling. Numbers of all baleen whales have declined to about 18% of former levels, the humpback and blue having suffered worst, being reduced to 3 and 5% of initial stocks. In terms of biomass, the baleen whales together represent some 35% of former levels. Sperm whales have not declined so greatly, now being about 50% of former biomass. Considering their migratory habits, Laws (1977) calculated the likely export of biomass from the Southern Ocean due to whales.

In the case of baleen whales, it is now only 14% of former levels, for the sperm whale 43% (though this estimate is based on incomplete data). The conclusion is that the reduction in stock of the baleen whales in particular has released a huge food resource, principally of euphausiids. At the same time, it has been noted that penguins and the fur and crabeater seals have increased in population size. These animals feed at least in part on euphausiids, and presumably have bene-fitted from the reduction in demand by baleen whales. Then not only has the reduction in baleen whale numbers had an effect on the growth and reproduc-tion of the remaining whales, but also on other species competing for the same food resources. The question remaining is whether the original balance could be regained, were this thought to be desirable. The uncertainty is increased in view of the proposals to develop krill fisheries in the Southern Ocean (Horwood, 1978).

9
Fishing and overfishing

Animal populations

In a given animal population there will be a number of individuals, this number being augmented by births (natality) and diminished by deaths (mortality). If both these processes are equal, or nearly so, the population size will be about constant. An excess of births leads to population growth, an excess of deaths to population decline. Sometimes populations do decline to local extinction, when the species ceases to be represented in a part of its former range. On the other hand they may grow, but not indefinitely: numbers usually stabilise about some upper limit which reflects the ability of the habitat to support the animal in question. A population which is more or less stable for several generations may be said to be at the carrying capacity for the habitat, that is, there is no further capacity to support further population growth. The notion of a population at carrying capacity, while difficult to quantify precisely, is evidently well founded. Although there is no such thing as an absolutely stable population and numbers do fluctuate, they do not fluctuate nearly as much as would be possible in the majority of species, given the rates of reproduction observed. It seems that there must be density-dependent processes at work which operate to regulate population size. The importance (or otherwise) of density dependent processes has been the subject of great controversy, which will not be reviewed here – a recent account may be found in Begon and Mortimer (1981). The position taken here is that density dependent processes are real and important in the regulation of natural populations. The evidence for this, together with a possible mechanism, was discussed in Chapter 8, for some fish examples.

Population growth

Fish population (or stock) growth may be modelled using one of the density-dependent growth equations. A full discussion of these can be found in May (1981), for our purposes a brief outline of essentials will suffice. A population of animals with several generations (such as most of the fish we have been interested in) will show what is called continuous growth, which may be described by the differential equation:

$$dN/dt = rN(1 - N/K). \qquad (9.1)$$

The population size is N, and the size at carrying capacity is K. The parameter r is an expression of the rate of increase when $N \ll K$. It is important to under-

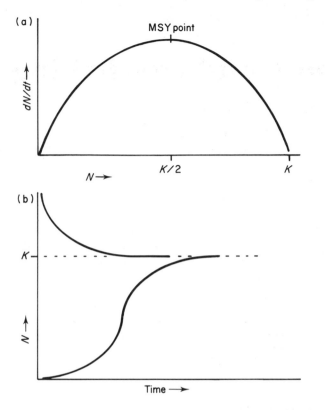

Fig. 9.1 **(a)** The relationship between *dN/dt* and *N* when *dN/dt* is described by an equation such as (9.1). The equilibrium population density is at *N* = *K*. (MSY = maximum sustainable yield, see text). **(b)** Growth of a population from below or above the carrying capacity density (*K*), described by equation (9.2).

stand that the expression *dN/dt* is to be read as a single symbol representing increase of *N* (population size) for unit *t* (time), that is, it describes the slope of the graph of *N* on *t*.

If this equation is evaluated to obtain values of *dN/dt* for different values of *N*, the graph would be as in Fig. 9.1a. This shows that at very small values of *N*, *dN/dt* is small, rising to a maximum at intermediate values of *N*, and falling again as *K* is approached. In fact, in the example given, *dN/dt* is maximal when *N* = *K*/2. That is, a population at half the carrying capacity is growing fastest. This is an estimate of what is called the maximum sustainable yield (MSY).

Equation 9.1, which gives *dN/dt*, can be converted to an equation for *N* by the mathematical process of integration. The result is:

$$N = \frac{K}{1 + [(K/N_0) - 1]\, \exp^{(-rt)}} \qquad (9.2)$$

and using equation 9.2 we can obtain successive estimates of *N* from an initial value. Such estimates can be used to plot a graph such as Fig. 9.1b, showing how a population will grow from low N or decline from high N towards K. In the first case, the familiar sigmoid curve of population growth is obtained.

Fishery regulation

Suppose that a fish stock is regulated in a way consistent with equation 9.1, showing density dependent logistic growth. If this stock is harvested, imposing additional mortality and reducing the size so that $N < K$, dN/dt initially increases, tending to bring N back to K. If however harvesting continues to depress the stock numbers, when N falls below the MSY point ($K/2$) dN/dt begins to decrease, and further exploitation is likely to push N rapidly down. This extremely simple model is the first step in trying to regulate a fishery, by defining the MSY point, catch (in terms of numbers of fish taken) can be regulated so that the stock is not depressed to values of N where dN/dt begins to fall.

An example of such regulation is shown graphically in Fig. 9.2. A harvest rate, h, can be defined as

$$h = qEN$$

where q is a catchability coefficient, set constant, E is fishing effort (e.g. boats per day) and N is stock size. The unharvested or pristine stock will grow at a rate defined by the logistic for $(dN/dt)_p$. Harvesting exerts extra mortality, so the harvested stock will grow at a rate defined by:

$$(dN/dt)_h = (dN/dt)_p - h.$$

It is useful to retain the expression dN/dt as a reminder that it is stock growth rate which is under consideration. Figure 9.2 shows the stock size (n) reached with a given moderate value of E, this is a stable value. Variations in E may shift the stock size to dangerously low values, but reduction in E in this event will allow recovery.

Maximum sustainable yield has been used widely as a basis for regulating harvest, and it has been widely unsuccessful. There are many reasons why this may have happened, and we shall look at two possibilities. In the first place, the model as so far described is biologically naive in that it takes no account of possible changes in behaviour with stock size or density. At low values of N school size may change, with more smaller schools. This tends to increase the mortality due to natural predation (Clark, 1981) and if this effect is large enough, the stock-recruitment relationship will change significantly. The nature of the change will be to disproportionately decrease $(dN/dt)_p$ at low values of N, so that the curve in Fig. 9.2 becomes like that in Fig.9.3a. Now a moderate value of E has two corresponding values of n, n_1 and n_2, and the first of these is unstable. If from an initially healthy state, effort rises to the so called 'critical case' (represented by the line E_c in Fig. 9.3a), the stock is depressed towards the unstable equilibrium n_1. Once this has happened, even moderate levels of E in future will not necessarily allow its recovery.

A similar situation may arise if the catchability coefficient, q, varies. Suppose that at low values of N, school size remains the same but number of schools is reduced and they occupy a smaller area. This may alter q increasing the catch-

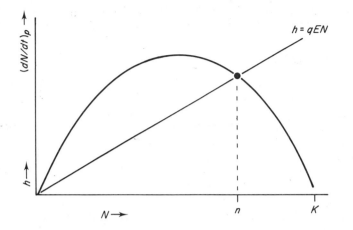

Fig. 9.2 The model described in equation (9.1) with moderate fishing incorporated. The new equilibrium population density will be at *n* instead of *K*. (After Clark, 1981.)

ability of the remaining fish. Now similarly to the case of depensation, there is a 'critical case' which may push a stock down to such a level that even with more moderate levels of *E*, it will not necessarily recover. This is illustrated in Fig. 9.3b as variable catchability.

These possibilities have been discussed by Clark (1981), who went further to point out that once economic factors begin to be introduced it becomes even more difficult to set sensible levels of harvest. While the economic factors are interesting, they are beyond the scope of this book, and the reader is referred to Clark's discussion. The very elementary discussion of stock growth and harvesting given here is of course only the most cursory introduction to the field of fishery mathematics, but is serves to show the basis for regulation. Discussing the development of fishery mathematics, May *et al.* (1978) considered that what is needed now is not further mathematical refinement, but for fishery practice to take account of inevitably fuzzy knowledge of stock levels and recruitment curves.

The history of the exploitation of fish and whale stocks is generally a sorry one. Figure 7.5 illustrates the point adequately: the Peru Current upwelling is more or less as productive as ever in terms of primary production, but the yield to man is much reduced following the collapse of the anchoveta fishery.

It will be clear from the brief discussion of stock growth and harvesting that regulation of fisheries depends on knowledge of the biology of the species, and often this knowledge is lacking. However we do know enough in the case of many species to set values for estimated yields that are better than guesses. It is also clear that regulation of the sort discussed is going to depend on the extent to which there is density-dependence in the stock-recruitment relationship. For fish which show low or very variable density-dependence, such as clupeoids, it becomes very difficult to forecast yields because of the often weak stock-recruitment relation. Moreover the school-forming clupeoids will be candidates for depensatory or variable catchability complications (Fig. 9.3). It is noteworthy that the clupeoids have proven to be notoriously difficult to manage, as

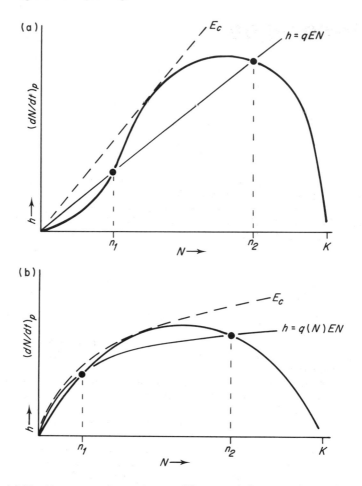

Fig. 9.3 (a) The 'depensatory' growth case. When population growth rate is reduced at low values of N there may exist a lower and unstable equilibrium population size (n_1) for a given level of fishing effort. Once a population has been pushed towards this value by fishing up to the critical case, the model predicts that it will not recover even with relaxation of effort. (b) With variation of the catchability coefficient (q) a similar situation may arise. (From Clark, 1981.)

is seen by the Peruvian anchoveta, the California sardine and anchovy, and the North Sea herring. All these fisheries have collapsed or shown major fluctuations this century.

The more are the complications, the more do the estimates of yield converge on the guess. In such circumstances, the 'safe' estimate of yield will be a low one, and this is often unattractive economically. Gaskin (1982) said of whales that '. . . there is no such thing as a whale population which will stand intense fishing pressure.' The history of fisheries shows the same to be true of some fish, although notably the gadoids and flatfish, with their high fecundity and relatively strong density-dependent stock-recruitment relationships, seem to be very resilient.

References

Alldredge, A.L. (1976). Field behaviour and adaptive strategies of appendicularians (Chordata: Tunicata). *Marine Biology, 38,* 29–39.

Alldredge, A.L. (1977). House morphology and mechanisms of feeding in the Oikopleuridae (Tunicata, Appendicularia). *Journal of Zoology, London, 181,* 175–88.

Anderson, L.W.J. and Sweeney, B.M. (1978). Role of inorganic ions in controlling sedimentation rate of a marine centric diatom *Ditylum brightwellii. Journal of Phycology,* 14, 204–14.

Assaf, G., Gerard, R. and Gordon A.L. (1971). Some mechanisms of oceanic mixing revealed in aerial photographs. *Journal of Geophysical Research,* 76, 6550–72.

Atkins, W.R.G. (1923). The phosphate content of fresh and salt water in its relation to the growth of algal plankton. Part 1. *Journal of the Marine Biological Association of the United Kingdom,* 13, 119–50.

Atkins, W.R.G. (1925). On the thermal stratification of sea water and its importance for the algal plankton. *Journal of the Marine Biological Association of the United Kingdom,* 13, 696–9.

Azam, F., Fenchel, T., Field, J.G., Gray, J.S. Meyer-Reil, L.A. and Thingstad, F. (1983). The ecological role of water-column microbes in the sea. *Marine Ecology Progress Series,* 10, 257–63.

Baalen, C. van and Brown, R.M. Jr. (1969). The ultrastructure of the marine blue green alga, *Trichodesmium erythraeum,* with special reference to the cell wall, gas vacuoles, and cylindrical bodies. *Archiv fur Mikrobiologie,* 69, 79–91.

Banse, K. (1976). Rates of growth, respiration and photosynthesis of unicellular algae as related to cell size – a review. *Journal of Phycology,* 12, 135–40.

Banse, K. (1982). Cell volumes, maximal growth rates of unicellular algae and ciliates, and the role of ciliates in the marine pelagial. *Limnology and Oceanography,* 27, 1059–71.

Barnett, T.P. (1977). An attempt to verify some theories of El Niño. *Journal of Physical Oceanography,* 7, 633–47.

Beers, J.R., Stevenson, M.R., Eppley, R.W. and Brooks, E.R. (1971). Plankton populations and upwelling off the coast of Peru, June 1969. *Fishery Bulletin,* 69, 859–76.

Begon, M. and Mortimer, M. (1981). *Population ecology: a unified study of animals and plants.* Blackwell Scientific Publications, Oxford.

Beklemishev, C.W. (1962). Superfluous feeding of marine herbivorous zooplankton. *Rapport et proces-verbaux des reunions: Conseil permanent international pour l'exploration de la mer,* 153, 108–13.

Belay, A. (1981). An experimental investigation of inhibition of phytoplankton photosynthesis at lake surfaces. *New Phytologist,* 89, 61–74.

Belay, A. and Fogg, G.E. (1978). Photoinhibition of photosynthesis in *Asterionella formosa* (Bacillariophyceae). *Journal of Phycology,* 14, 341–7.

Bell, W.H. (1983). Bacterial utilization of algal extracelluar products. 3. The specificity of algal-bacterial interaction. *Limnology and Oceanography,* 28, 1131–43.

Berman, T. and Holm-Hansen, O. (1974). Release of photoassimilated carbon as dissolved organic matter by marine phytoplankton. *Marine Biology,* **28**, 305-10.

Bienfang, P.K. (1980). Phytoplankton sinking rates in oligotrophic waters off Hawaii, USA. *Marine Biology,* **61**, 69-77.

Bienfang, P.K., Harrison P.J. and Quarmby, L.M. (1982). Sinking rate response to depletion of nitrate, phosphate and silicate in four marine diatoms. *Marine Biology,* **67**, 295-302.

Blackburn, M. (1973). Regressions between biological oceanographic measurements in the eastern tropical Pacific and their significance to ecological efficiency. *Limnology and Oceanography,* **18**, 552-63.

Blackburn, M. (1981). Low latitude gyral regions. In: *Analysis of marine ecosystems.* Longhurst, A.R. (ed.). Academic Press, London.

Blaxter, J.H.S. and Hunter, J.R. (1982). The biology of the clupeoid fishes. *Advances in Marine Biology,* **20**, 1-223.

Bonner, W.N. (1980). *Whales.* Blandford Press, Poole, Dorset.

Bothe, H. (1982). Nitrogen fixation. In: *The biology of Cyanobacteria,* Carr, N.G. and Whitton, B.A. (eds). Blackwell Scientific Publications, Oxford.

Boyd, C.M. (1976). Selection of particle sizes by filter-feeding copepods: a plea for reason. *Limnology and Oceanography,* **21**, 175-80.

Bricaud, A., Morel, A. and Prieur, L. (1981). Absorption by dissolved organic matter of the sea (yellow substance) in the UV and visible domains. *Limnology and Oceanography,* **26**, 43-53.

Brodie, P.F. (1975). Cetacean energetics, an overview of intraspecific size variation. *Ecology,* **56**, 152-61.

Brody, M. and Brody, S.S. (1962). Light reactions in photosynthesis. In: *Physiology and biochemistry of algae.* Lewin, R.A. (ed.). Academic Press, New York.

Brown, S.G. and Lockyer, C.H. (1984). Whales. In: *Antarctic ecology.* Volume 2. Laws, R.M. (ed.). Academic Press, London.

Butler, W.L. (1978). Energy distribution in the photochemical apparatus of photosynthesis. *Annual Review of Plant Physiology,* **29**, 345-78.

Cannon, H.G. (1928). On the feeding mechanism of the copepods, *Calanus finmarchicus* and *Diaptomus gracilis. British Journal of Experimental Biology,* **6**, 131-44.

Cannon, H.G. and Manton, S.M. (1927). On the feeding mechanism of a mysid crustacean, *Hemimysis lamornae. Transactions of the Royal Society of Edinburgh,* **55**, 219-53.

Carpenter, E.J. and Price, C.C. IV. (1976). Marine *Oscillatoria (Trichodesmium):* explanation for aerobic nitrogen fixation without heterocysts. *Science Washington,* **191**, 1278-80.

Carpenter, E.J. and Price, C.C. IV. (1977). Nitrogen fixation, distribution, and production of *Oscillatoria (Trichodesmium)* spp. in the western Sargasso and Caribbean seas. *Limnology and Oceanography,* **22**, 60-72.

Chan, A.T. (1980). Comparative physiological study of marine diatoms and dinoflagellates in relation to irradiance and cell size. II. Relationship between photosynthesis, growth, and carbon/chlorophyll *a* ratio. *Journal of Phycology,* **16**, 428-32.

Clark, C.W. (1981). Bioeconomics. In: *Theoretical ecology: principles and applications.* Second edition. May, R.M. (ed.). Blackwell Scientific Publications, Oxford.

Clarke, A. (1985). Energy flow in the Southern Ocean food web. In: *Antarctic nutrient cycles and food webs.* Siegfried, W.R., Condy, P.R. and Laws, R.M. (eds). Springer-Verlag, Berlin.

Conover, R.J. (1956). Oceanography of Long Island Sound, 1952-1954. VI. Biology of *Acartia clausi* and *A. tonsa. Bulletin of the Bingham Oceanographic Collection,* **15**, 156-223.

Conover, R.J. (1978). Feeding interactions in the pelagic zone. *Rapport et procesverbaux des reunions: Conseil permanent international pour l'exploration de la mer,* **173**, 66-76.

Cooper, L.H.N. (1938). Phosphate in the English Channel, 1933–8, with a comparison with earlier years, 1916 and 1923–32. *Journal of the Marine Biological Association of the United Kingdom,* **23,** 181–95.

Cooper, L.H.N. (1948). Phosphate and fisheries. *Journal of the Marine Biological Association of the United Kingdom,* **27,** 326–36.

Corner, E.D.S., Head, R.N., Kilvington, C.C. and Marshall, S.M. (1974). On the nutrition and metabolism of zooplankton. IX. Studies relating to the nutrition of overwintering *Calanus. Journal of the Marine Biological Association of the United Kingdom,* **54,** 319–31.

Cowles, T.J., Barber, R.T. and Guillen, O. (1977). Biological consequences of the 1975 El Niño. *Science, Washington,* **195,** 285–7.

Cullen, J.J. and Horrigan, S.G. (1981). Effects of nitrate on the diurnal vertical migration, carbon to nitrogen ratio, and photosynthetic capacity of the dinoflagellate *Gymnodinium splendens. Marine Biology,* **62,** 81–9.

Currie, D.J. (1984). Microscale nutrient patches: do they matter to the phytoplankton? *Limnology and Oceanography,* **29,** 211–14.

Cushing, D.H. (1967). The grouping of herring populations. *Journal of the Marine Biological Association of the United Kingdom,* **47,** 193–208.

Cushing, D.H. (1971). Upwelling and the production of fish. *Advances in Marine Biology,* **9,** 225–334.

Cushing, D.H. (1973a). Dependence of recruitment on parent stock. *Journal of the Fisheries Research Board of Canada,* **30,** 1965–76.

Cushing, D.H. (1973b). Production in the Indian Ocean and the transfer from the primary to the secondary level. In: *The biology of the Indian Ocean.* Zeitzschel, B. and Gerlach, S.A. (eds). Springer-Verlag, Berlin.

Cushing, D.H. (1975). *Marine ecology and fisheries.* Cambridge University Press, Cambridge.

Cushing, D.H. (1976). The impact of climatic change on fish stocks in the north Atlantic. *Geographical Journal,* **142,** 216–27.

Cushing, D.H. (1980). The decline of the herring stocks and the gadoid outburst. *Journal du Conseil: Conseil permanent international pour l'exploration de la mer,* **39,** 70–81.

Cushing, D.H. (1982). *Climate and fisheries.* Academic Press, London.

Cushing, D.H. and Harris, J.G.K. (1973). Stock and recruitment and the problem of density dependence. *Rapport et proces-verbaux des reunions: Conseil permanent international pour l'exploration de la mer,* **164,** 142–55.

Dagg, M. (1977). Some effects of patchy food environments on copepods. *Limnology and Oceanography,* **22,** 99–107.

Darley, W.M. and Volcani, B.E. (1969). Role of silicon in diatom metabolism. A silicon requirement for deoxyribonucleic acid synthesis in the diatom *Cylindrotheca fusiformis* Reimann and Lewin. *Experimental Cell Research,* **58,** 334–42.

DeAngelis, D.L. (1980). Energy flow, nutrient cycling, and ecosystem resilience. *Ecology,* **61,** 764–71.

Defant, A. (1961). *Physical Oceanography.* Vol. l. Pergamon Press, Oxford.

Deibel, D. (1986). Feeding mechanism and house of the appendicularian *Oikopleura vanhoeffeni. Marine Biology,* **93,** 429–36.

Denman, K.L. and Platt, K.T. (1976). The variance spectrum of phytoplankton in a turbulent ocean. *Journal of Marine Research,* **34,** 593–601.

Dodge, J.D. and Hart-Jones, B. (1974). The vertical and seasonal distribution of dinoflagellates in the North Sea. *Botanica Marina,* **17,** 113–17.

Dugdale, R.C. (1967). Nutrient limitation in the sea: dynamics, identification, and significance. *Limnology and Oceanography,* **12,** 685–95.

Elton, C. (1966). *Animal ecology.* Methuen and Co. Ltd., London.

Eppley, R.W., Holm-Hansen, O. and Strickland, J.D.H. (1968). Some observations on the vertical migration of dinoflagellates. *Journal of Phycology,* **4**, 333–40.

Eppley, R.W., Rogers, J.N. and McCarthy, J.J. (1969). Half-saturation constants for uptake of nitrate and ammonium by marine phytoplankton. *Limnology and Oceanography,* **14**, 912–20.

Eppley, R.W. and Thomas, W.H. (1969). Comparison of half-saturation constants for growth and nitrate uptake of marine phytoplankton. *Journal of Phycology,* **5**, 375–9.

Esaias, W.E. and Curl, H.C. Jr. (1972). Effect of dinoflagellate bioluminescence on copepod ingestion rates. *Limnology and Oceanography,* **17**, 901–6.

Esterly, C.O. (1916). The feeding habits and food of pelagic copepods and the question of nutrition by organic substances in solution in the water. *University of California publications in Zoology,* **16**, 171–84.

Everson, I. (1984). Marine interactions. In: *Antarctic ecology.* Volume 2. Laws, R.M. (ed.). Academic Press, London.

Falkowski, P.G. (1983). Light-shade adaptation and vertical mixing of marine phytoplankton: a comparative field study. *Journal of Marine Research,* **41**, 215–37.

Faller, A.J. (1969). The generation of Langmuir circulations by the eddy pressure of surface waves. *Limnology and Oceanography,* **14**, 504–13.

Faller, A.J. and Caponi, E.A. (1978). Laboratory studies of wind-driven Langmuir circulations. *Journal of Geophysical Research,* **83**, 3617–33.

Feigenbaum, D.L. and Maris, R.C. (1984). Feeding in the Chaetognatha. *Oceanography and Marine Biology Annual Review,* **22**, 343–92.

Fenchel, T. (1982). Ecology of heterotrophic microflagellates. I. Some important forms and their functional morphology. *Marine Ecology Progress Series,* **8**, 211–23.

Fiedler, P.C. (1982). Zooplankton avoidance and reduced grazing responses to *Gymnodinium splendens (Dinophyceae). Limnology and Oceanography,* **27**, 961–5.

Fogg, G.E. (1966). The extracellular products of algae. *Oceanography and Marine Biology Annual Review,* **4**, 195–212.

Fogg, G.E. (1982). Marine plankton. In: *The biology of the Cyanobacteria.* Botanical Monographs, Volume 19. Carr, N.G. and Whitton, B.A. (eds). Blackwell Scientific Publications, Oxford.

Foulds, J.B. and Roff, J.C. (1976). Oxygen consumption during simulated vertical migration in *Mysis relicta* (Crustacea, Mysidacea). *Canadian Journal of Zoology,* **54**, 377–85.

Friebele, E.S. Correll, D.L. and Faust, M.A. (1978). Relationship between phytoplankton cell size and the rate of orthophosphate uptake: *in situ* observations of an estuarine population. *Marine Biology,* **45**, 39–52.

Friedman, M.M. and Strickler, J.R. (1975). Chemoreceptors and feeding in calanoid copepods (Arthropoda: Crustacea). *Proceedings of the National Academy of Sciences USA,* **72**, 4185–8.

Frost, B.W. (1972). Effects of size and concentration of food particles on the feeding behavior of the marine planktonic copepod *Calanus pacificus. Limnology and Oceanography,* **17**, 805–15.

Frost, B.W. (1975). A threshold feeding behavior in *Calanus pacificus. Limnology and Oceanography,* **20**, 263–6.

Frost, B.W. (1977). Feeding behavior of *Calanus pacificus* in mixtures of food particles. *Limnology and Oceanography,* **22**, 472–91.

Fuller, J.L. (1937). Feeding rate of *Calanus finmarchicus* in relation to environmental conditions. *Biological Bulletin,* **72**, 233–46.

Garrod, D.J. and Jones, B.W. (1974). Stock and recruitment relationship in the northeast Arctic cod stock and the implications for management of the stock. *Journal du Conseil: Conseil permanent international pour l'exploration de la mer,* **36**, 35–41.

Gaskin, D.E. (1982). *The ecology of whales and dolphins.* Heinemann, London.

Gates, D.M. (1962). *Energy exchange in the biosphere.* Harper & Row, New York.

Gaudy, R. (1974). Feeding four species of pelagic copepods under experimental conditions. *Marine Biology,* **25,** 125–41.

Gauld, D.T. (1951). The grazing rate of planktonic copepods. *Journal of the Marine Biological Association of the United Kingdom,* **31,** 461–74.

Gauld, D.T. (1964). Feeding in planktonic copepods. In: *Grazing in terrestrial and marine environments.* Crisp, D.J. (ed.). Blackwell Scientific Publications, Oxford.

Gilmer, R.W. (1972). Free-floating mucus webs: a novel feeding adaptation for the open ocean. *Science, Washington,* **176,** 1239–40.

Gilmer, R.W. (1974). Some aspects of feeding in thecosomatous pteropod molluscs. *Journal of experimental Marine Biology and Ecology,* **15,** 127–44.

Goldman, J.C. (1976). Identification of nitrogen as a growth-limiting nutrient in wastewaters and coastal marine waters through continuous culture algal assays. *Water Research,* **10,** 97–104.

Goldman, J.C. (1984). Oceanic nutrient cycles. In: *Flows of energy and materials in marine ecosystems: theory and practice.* Fasham, M.J.R. (ed.). Plenum Press, New York.

Grahame, J. (1976). Zooplankton of a tropical harbour: the numbers, composition, and response to physical factors of zooplankton in Kingston harbour, Jamaica. *Journal of experimental marine Biology and Ecology,* **25,** 219–37.

Grahame, J. (1983). Adaptive aspects of feeding mechanisms. In: *The Biology of Crustacea.* Volume 8. Vernberg, J.F. and Vernberg, W.B. (eds). Academic Press, New York.

Grahame, J., and Branch, G. (1985). Reproductive patterns of marine invertebrates. *Oceanography and Marine Biology Annual Review,* **23,** 373–98.

Guillen, O.G. and Calienes, R.Z. (1981). Biological productivity and El Niño. In: *Resource management and environmental uncertainty: lessons from coastal upwelling fisheries.* Glantz, M.H. and Thompson, J.D. (eds). Wiley Interscience, New York.

Gulland, J.A. and Garcia, S. (1984). Observed patterns in multispecies fisheries. In: *Exploitation of marine communities.* May, R.M. (ed.). Springer-Verlag, Berlin.

Gyllenberg, G. and Lundqvist, G. (1978). Utilization of dissolved glucose by two copepod species. *Annales Zoologici Fennici,* **15,** 323–7.

Hand, W.G., Collard, P.A. and Davenport, D. (1965). The effects of temperature and salinity change on swimming rate in the dinoflagellates, Gonyaulax and Gymnodinium. *Biological Bulletin,* **128,** 90–101.

Hardy, A.C. (1924). The herring in relation to its animate environment. Part I. The food and feeding habits of the herring with special reference to the east coast of England. *Fishery Investigations, Series II.* Volume VII, number 3, 53 pp.

Hardy, A.C. (1953). Some problems of pelagic life. In: *Essays in marine biology.* Marshall, S.M. and Orr, A.P. (eds). Oliver & Boyd, Edinburgh.

Harris, G.P. (1980). The measurement of photosynthesis in natural populations of phytoplankton. In: *The physiological ecology of phytoplankton.* Morris, I. (ed.). Blackwell Scientific Publications, Oxford.

Harrison, R.J. (1969). Reproduction and reproductive organs. In: *The biology of marine mammals.* Andersen, H.T. (ed.). Academic Press, New York.

Harrison, W.G. (1980). Nutrient regeneration and primary production in the sea. In: *Primary productivity in the sea.* Falkowski, P.G. (ed.). Plenum Press, New York.

Hart, T.J. and Currie, R.I. (1960). The Benguela current. *Discovery Reports,* **31,** 123–298.

Harvey, H.W. (1925). Evaporation and temperature changes in the English Channel. *Journal of the Marine Biological Association of the United Kingdom,* **13,** 678–92.

Harvey, H.W. (1937). Note on selective feeding by Calanus. *Journal of the Marine Biological Association of the United Kingdom,* **22,** 97–100.

Harvey, H.W. (1950). On the production of living matter in the sea off Plymouth.

Journal of the Marine Biological Association of the United Kingdom, **29**, 97–137.

Harvey, H.W. (1955). *The chemistry and fertility of sea waters.* Cambridge University Press, London.

Harvey, H.W., Cooper, L.H.N., Lebour, M.V. and Russell, F.S. (1935). The control of production. *Journal of the Marine Biological Association of the United Kingdom*, **20**, 407 41.

Heinle, D.R., Harris, R.P., Ustach, J.F. and Flemer, D.A. (1977). Detritus as food for estuarine copepods. *Marine Biology*, **40**, 341–53.

Holligan, P.M. and Harbour, D.S. (1977). The vertical distribution and succession of phytoplankton in the western English Channel in 1975 and 1976. *Journal of the Marine Biological Association of the United Kingdom*, **57**, 1075–93.

Holligan, P.M., Harris, R.P., Newell, R.C., Harbour, D.S., Head, R.N., Linley, E.A.S., Lucas, M.I., Tranter, P.R.G. and Weekley, C.M. (1984a). Vertical distribution and partitioning of organic carbon in mixed, frontal and stratified waters of the English Channel. *Marine Ecology Progress Series*, **14**, 111–27.

Holligan, P.M., Williams, P.J.leB., Purdie, D. and Harris, R.P. (1984b). Photosynthesis, respiration and nitrogen supply of plankton populations in stratified, frontal and tidally mixed shelf waters. *Marine Ecology Progress Series*, **17**, 201–13.

Horwood, J.W. (1978). Whale management and the potential fishery for krill. *Report. International Whaling Commission*, **28**, 187–9.

Horwood, J.W. (1980). Population biology and stock assessment of southern hemisphere sei whales. *Report. International Whaling Commission*, **30**, 519–30.

Isaacs, J.D., Tont, S.A. and Wick, G.L. (1974). Deep scattering layers: vertical migration as a tactic for finding food. *Deep-Sea Research*, **21**, 651–6.

Jenkins, J.T. (1925). *The fishes of the British Isles, both fresh water and salt.* Frederick Warne & Co. Ltd., London.

Jennings, J.C. Jr., Gordon, L.I. and Nelson, D.M. (1984). Nutrient depletion indicates high primary productivity in the Weddell sea. *Nature, London*, **309**, 51–4.

Jerlov, N.G. (1968). *Optical oceanography.* Elsevier Publishing Company, Amsterdam.

Johnson, P.W. and Sieburth, J.McN. (1979). Chroococcoid cyanobacteria in the sea: a ubiquitous and diverse phototrophic biomass. *Limnology and Oceanography*, **24**, 928–35.

Jørgensen, C.B. (1976). August Putter, August Krogh, and modern ideas on the use of dissolved organic matter in aquatic environments. *Biological Reviews*, **51**, 291–328,

Jørgensen, E.G. (1977). Photosynthesis. In: *The biology of diatoms.* Botanical Monographs, vol. 13. Werner, E. (ed.). Blackwell Scientific Publications, Oxford.

Kahn, N. and Swift, E. (1978). Positive buoyancy through ionic control in the nonmotile marine dinoflagellate *Pyrocystis noctiluca* Murray ex Schuett. *Limnology and Oceanography*, **23**, 649–58.

King, K.R., Hollibaugh, J.T. and Azam, F. (1980). Predator-prey interactions between the larvacean *Oikopleura dioica* and bacterioplankton in enclosed water columns. *Marine Biology*, **56**, 49–57.

Kirk, J.T.O. (1983). *Light and photosynthesis in aquatic ecosystems.* Cambridge University Press, Cambridge.

Koehl, M.A.R. and Strickler, J.R. (1981). Copepod feeding currents: food capture at low Reynolds number. *Limnology and Oceanography*, **26**, 1062–73.

LaBarbera, M. (1984). Feeding currents and particle capture mechanisms in suspension feeding animals. *American Zoologist*, **24**, 71–84.

Lancelot, C. and Billen, G. (1984). Activity of heterotrophic bacteria and its coupling to primary production during the spring phytoplankton bloom in the southern bight of the North Sea. *Limnology and Oceanography*, **29**, 721–30.

Langmuir, I. (1938). Surface motion of water induced by wind. *Science, Washington*, **87**, 119–23.

Laws, E.A., Redalje, D.G., Haas, L.W., Bienfang, P.K., Eppley, R.W., Harrison, W.G., Karl, D.M. and Marra, J. (1984). High phytoplankton growth and production rates in oligotrophic Hawaiian coastal waters. *Limnology and Oceanography,* **29,** 1161-9.

Laws, R.M. (1977). Seals and whales of the Southern ocean. *Philosophical Transactions of the Royal Society of London,* **B279,** 81-96.

Lehman, J.T. (1977). On calculating drag characteristics for decelerating zooplankton. *Limnology and Oceanography,* **22,** 170-72.

Lehman, J.T. and Scavia, D. (1984). Measuring the ecological significance of microscale nutrient patches. *Limnology and Oceanography,* **29,** 214-16.

Lewis, M.R., Horne, E.P.W., Cullen, J.J., Oakey, N.S. and Platt, T. (1984). Turbulent motions may control phytoplankton photosynthesis in the upper ocean. *Nature, London,* **311,** 49-50.

Lieth, H. (1975). Primary productivity in ecosystems: comparative analysis of global patterns. In: *Unifying concepts in ecology.* Dobben, W.H. van and Lowe-McConnell, R.H. (eds). Dr W. Junk B.V. Publishers, The Hague.

Lillelund, K. and Lasker, R. (1971). Laboratory studies on predation by marine copepods on fish larvae. *Fishery Bulletin,* **69,** 655-67.

Longhurst, A.R. (1976). Vertical migration. In: *The ecology of the seas.* Cushing, D.H. and Walsh, J.J. (eds). Blackwell Scientific Publications, Oxford.

McCarthy, J.J. (1980). Nitrogen. In: *The physiological ecology of phytoplankton.* Morris, I. (ed.). Blackwell Scientific Publications, Oxford.

McCarthy, J.J. and Carpenter, F.J. (1979). *Oscillatoria (Trichodesmium) thiebautii* (Cyanophyta) in the central north Atlantic ocean. *Journal of Phycology,* **15,** 75-82.

McCarthy, J.J. and Goldman, J.C. (1979). Nitrogenous nutrition of marine phytoplankton in nutrient-depleted waters. *Science, Washington,* **203,** 670-2.

McCarthy, J.J., Taylor, W.R. and Loftus, M.E. (1974). Significance of nanoplankton in the Chesapeake Bay estuary and problems associated with the measurement of nanoplankton productivity. *Marine Biology,* **24,** 7-16.

McGowan, J.A. and Hayward, T.L. (1978). Mixing and oceanic productivity. *Deep-Sea Research,* **25,** 771-93.

MacIsaac, J.J. and Dugdale, R.C. (1969). The kinetics of nitrate and ammonia uptake by natural populations of marine phytoplankton. *Deep-Sea Research,* **16,** 45-57.

Magre, T.H. Weare, N.M. and Holm-Hansen, O. (1974). Nitrogen fixation in the north Pacific ocean. *Marine Biology,* **24,** 109-19.

Malone, T.C. (1980). Algal size. In: *The physiological ecology of phytoplankton.* Morris, I. (ed.). Blackwell Scientific Publications, Oxford.

Malone, T.C. (1982). Phytoplankton photosynthesis and carbon-specific growth: light saturated rates in a nutrient-rich environment. *Limnology and Oceanography,* **27,** 226-35.

Malone, T.C., Falkowski, P.G., Hopkins, T.S., Rowe, G.T. and Whitledge, T.E. (1983). Mesoscale response of diatom populations to a wind event in the plume of the Hudson River. *Deep-Sea Research,* **30,** 149-70.

Margalef, R. (1958). Temporal succession and spatial heterogeneity in phytoplankton. In: *Perspectives in marine biology.* Buzzati-Traverso, A.A. (ed.). University of California Press, Berkeley.

Margalef, R. (1967). The food web in the pelagic environment. *Helgolander Wissenschaftliche Meeresuntersuchungen,* **15,** 548-59.

Margalef, R. (1978). Life-forms of phytoplankton as survival alternatives in an unstable environment. *Oceanologica Acta,* **1,** 493-509.

Marshall, S.M. and Orr, A.P. (1930). A study of the spring diatom increase in Loch Striven. *Journal of the Marine Biological Association of the United Kingdom,* **16,** 853-78.

Mauchline, J. (1980). The biology of mysids and euphausiids. *Advances in Marine Biology,* **18**.

Mauchline, J. and Fisher, L.R. (1969). The biology of euphausiids. *Advances in Marine Biology,* **7**.

May, R.M. (1981). Models for single populations. In: *Theoretical ecology: principles and applications* Second edition. May, R.M. (ed.). Blackwell Scientific Publications, Oxford.

May, R.M., Beddington, J.R., Horwood, J.W. and Shepherd, J.G. (1978). Exploiting natural populations in an uncertain world. *Mathematical Biosciences,* **42**, 219-52.

Mayzaud, P. and Poulet, S.A. (1978). The importance of the time factor in the response of zooplankton to varying concentrations of naturally occurring particulate matter. *Limnology and Oceanography,* **23**, 1144-54.

Monahan, E.C. and Pybus, M.J. (1978). Colour, ultraviolet absorbance and salinity of the surface waters off the west coast of Ireland. *Nature, London,* **274**, 782-4.

Morel, A. and Prieur, L. (1977). Analysis of variations in ocean color. *Limnology and Oceanography,* **22**, 709-22.

Morel, A. and Smith, R.C. (1974). Relation between total quanta and total energy for aquatic photosynthesis. *Limnology and Oceanography,* **19**, 591-600.

Morton, J.E. (1954). The biology of *Limacina retroversa. Journal of the marine biological Association of the United Kingdom,* **33**, 297-312.

Mullin, M.M. and Brooks, E.R. (1976). Some consequences of distributional heterogeneity of phytoplankton and zooplankton. *Limnology and Oceanography,* **21**, 784-96.

Murphy, G.I. (1968). Pattern in life history and the environment. *American Naturalist,* **102**, 391-403.

Murphy, G.I. (1977). Clupeoids. In: *Fish population dynamics.* Gulland, J.A. (ed.). John Wiley & Sons, London.

Nemoto, T. and Harrison, G. (1981). High latitude ecosystems. In: *Analysis of marine ecosystems.* Longhurst, A.R. (ed.). Academic Press, London.

Newell, R.C. and Linley, E.A.S. (1984). Significance of microheterotrophs in the decomposition of phytoplankton: estimates of carbon and nitrogen flow based on the biomass of plankton communities. *Marine Ecology Progress Series,* **16**, 105-19.

Nicholls, A.G. (1933). On the biology of *Calanus finmarchicus.* III: Vertical distribution and vertical migration in the Clyde Sea area. *Journal of the Marine Biological Association of the United Kingdom,* **19**, 139-64.

Norris, K. and Mohl, B. (1983). Can odontocetes debilitate prey with sound? *American Naturalist,* **122**, 85-104.

Odum, E.P. (1983). *Basic ecology.* Saunders College Publishing, Philadelphia.

Omori, M. (1974). The biology of pelagic shrimps in the ocean. *Advances in Marine Biology,* **12**, 233-324.

Paasche, E. (1980). Silicon. In: *The physiological ecology of phytoplankton.* Morris, I. (ed.). Blackwell Scientific Publications, Oxford.

Paffenhöfer, G.-A. (1973). The cultivation of an appendicularian through numerous generations. *Marine Biology,* **22**, 183-5.

Paffenhöfer, G.-A., Strickler, J.R. and Alcaraz, M. (1982). Suspension-feeding by herbivorous calanoid copepods: a cinematographic study. *Marine Biology,* **67**, 193-9.

Parke, M. and Dixon, P.S. (1976). Check-list of British marine algae – third revision. *Journal of the Marine Biological Association of the United Kingdom,* **56**, 527-94.

Parrish, B.B. and Saville, A. (1965). The biology of the north-east Atlantic herring populations. *Oceanography and Marine Biology Annual Review,* **3**, 323-73.

Parsons, T.R., LeBrasseur, R.J. and Fulton, J.D. (1967). Some observations on the dependence of zooplankton grazing on the cell size and composition of phytoplankton blooms. *Journal of the Oceanographical Society of Japan,* **23**, 10-17.

Parsons, T.R. and Takahashi, M. (1973). Environmental control of phytoplankton cell size. *Limnology and Oceanography,* **18**, 511-24.

Parsons, T.R. and Takahashi, M. (1974). A rebuttal to the comment by Hecky and Kilham. *Limnology and Oceanography,* **19**, 366-8.

Peters, R.H. (1983). *The ecological implications of body size.* Cambridge University Press, Cambridge.

Peterson, B.J. (1980). Aquatic primary productivity and the $^{14}CO-C_2$ method: a history of the productivity problem. *Annual Review of Ecology and Systematics,* **11**, 359-85.

Peterson, W.T., Miller, C.B. and Hutchinson, A. (1979). Zonation and maintenance of copepod populations in the Oregon upwelling zone. *Deep-Sea Research,* **26**, 467-94.

Philander, S.G.H. (1983). El Niño southern oscillation phenomena. *Nature, London,* **302**, 295-301.

Pickard, G.L. and Emery, W.J. (1982). *Descriptive physical oceanography: an introduction.* Pergamon Press, Oxford.

Pimm, S.L. (1982). *Food webs.* Chapman & Hall, London.

Pimm, S.L. and Lawton, J.H. (1977). Number of trophic levels in ecological communities. *Nature, London,* **268**, 329-31.

Pingree, R.D. (1978). Cyclonic eddies and cross-frontal mixing. *Journal of the Marine Biological Association of the United Kingdom,* **58**, 955-63.

Pingree, R.D., Holligan, P.M., Mardell, G.T. and Head, R.N. (1976). The influence of physical stability on spring, summer and autumn phytoplankton blooms in the Celtic sea. *Journal of the Marine Biological Association of the United Kingdom,* **56**, 845-73.

Pingree, R.D., Maddock, L. and Butler, E.I. (1977). The influence of biological activity and physical stability in determining the chemical distributions of inorganic phosphate, silicate and nitrate. *Journal of the Marine Biological Association of the United Kingdom,* **57**, 1065-73.

Platt, T. and Denman, K. (1980). Patchiness in phytoplankton distribution. In: *The physiological ecology of phytoplankton.* Morris, I. (ed.). Blackwell Scientific Publications, Oxford.

Platt, T., Rao, D.V.S. and Irwin, B. (1983). Photosynthesis of picoplankton in the oligotrophic ocean. *Nature, London,* **301**, 702-4.

Poulet, S.A. (1976). Feeding of *Pseudocalanus minutus* on living and non-living particles. *Marine Biology,* **34**, 117-25.

Poulet, S.A. and Marsot, P. (1978). Chemosensory grazing by marine calanoid copepods (Arthropoda: Crustacea). *Science, Washington,* **200**, 1403-5.

Raymont, J.E.G. (1980). *Plankton and productivity in the oceans, second edition. Volume 1, Phytoplankton.* Pergamon, Oxford.

Raymont, J.E.G. (1983). *Plankton and productivity in the oceans, second edition. Volume 2, Zooplankton.* Pergamon, Oxford.

Rigler, F.H. (1975). The concept of energy flow and nutrient flow between trophic levels. In: *Unifying concepts in ecology.* Dobben, W.H. van and Lowe-McConnel, R.H. (eds). Dr W. Junk B.V. Publishers, The Hague.

Riley, G.A. (1967). The plankton of estuaries. In: *Estuaries.* Lauff, G.H. (ed.). American Association for the Advancement of Science, Washington.

Ring Group. (1981). Gulf Stream cold-core rings: their physics, chemistry and biology. *Science, Washington,* **212**, 1091-100.

Roff, D.A. (1981). Reproductive uncertainty and the evolution of iteroparity: why don't flatfish put all their eggs in one basket? *Canadian Journal of Fisheries and Aquatic Sciences,* **38**, 968-77.

Roger, G.C. and Grandperrin, R. (1976). Pelagic food webs in the tropical Pacific. *Limnology and Oceanography,* **21**, 731-5.

Roman, M.R. (1978). Ingestion of the blue-green alga *Trichodesmium* by the harpacticoid copepod, *Macrosetella gracilis. Limnology and Oceanography,* **23**, 1245-8.

Ryther, J.H. (1969). Photosynthesis and fish production in the sea. *Science, Washington,* **166**, 72–6.

Ryther, J.H., Menzel, D.W., Hulburt, E.M., Lorenzen, C.J. and Corwin, N. (1971). The production and utilization of organic matter in the Peru coastal current. *Investigacion Pesquera,* **35**, 43–59.

Ryther, J.H and Yentsch, C.S. (1958). Primary production of continental shelf waters off New York. *Limnology and Oceanography,* **3**, 327–35.

Sameoto, D.D. (1972). Yearly respiration rate and estimated energy budget for *Sagitta elegans. Journal of the Fisheries Research Board of Canada,* **29**, 987–96.

Sameoto, D.D. (1973). Annual life cycle and production of the chaetognath *Sagitta elegans* in Bedford Basin, Nova Scotia. *Journal of the Fisheries Research Board of Canada,* **30**, 333–44.

Scott, J.T., Myer, G.E., Stewart, R. and Walther, E.G. (1969). On the mechanism of Langmuir circulations and their role in epilimnion mixing. *Limnology and Oceanography,* **14**, 493–503.

Semina, H.J. (1972). The size of phytoplankton cells in the Pacific ocean. *Internationale Revue der gesamten Hydrobiologie u Hydrographie,* **57**, 177–205.

Sharp, J.H. (1977). Excretion of organic matter by marine phytoplankton: do healthy cells do it? *Limnology and Oceanography,* **22**, 381–99.

Sheader, M. and Evans, F. (1975). Feeding and gut structure of *Parathemisto gaudichaudi* (Guerin) (Amphipoda, Hyperiidea). *Journal of the Marine Biological Association of the United Kingdom,* **55**, 641–56.

Shelbourne, J.E. (1953). The feeding habits of plaice post-larvae in the southern bight. *Journal of the Marine Biological Association of the United Kingdom,* **32**, 149–59.

Shelbourne, J.E. (1957). The feeding and condition of plaice larvae in good and bad plankton patches. *Journal of the Marine Biological Association of the United Kingdom,* **36**, 539–52.

Sheldon, R.W. (1984). Phytoplankton growth rates in the tropical ocean. *Limnology and Oceanography,* **29**, 1342–6.

Shepherd, J.G. and Cushing, D.H. (1980). A mechanism for density-dependent survival of larval fish as the basis of a stock-recruitment relationship. *Journal du Conseil: Conseil permanent international pour l'exploration de la mer,* **39**, 160–7.

Shimura, S. and Fujita, Y. (1975). Phycoerythrin and photosynthesis of the pelagic blue-green alga *Trichodesmium thiebautii* in the waters of Kuroshio, Japan. *Marine Biology,* **31**, 121–8.

Sieburth, J. McN. and Jensen, A. (1968). Studies on algal humic substances in the sea. I. Gelbstoff (humic material) in terrestrial and marine waters. *Journal of experimental marine Biology and Ecology,* **2**, 174–89.

Slijper, E.J. (1962). *Whales.* Hutchinson, London.

Slobodkin, L. (1962). Energy in animal ecology. *Advances in Ecological Research,* **1**, 69–101.

Smayda, T.J. (1970). The suspension and sinking of phytoplankton in the sea. *Oceanography and Marine Biology Annual Review,* **8**, 353–414.

Smayda, T.J. (1980). Phytoplankton species succession. In: *The physiological ecology of phytoplankton.* Morris, I. (ed.). Blackwell Scientific Publications, Oxford.

Smith, W.O., Heburn, G.W., Barber, R.T. and O'Brien, J.J. (1983). Regulation of phytoplankton communities by physical processes in upwelling ecosystems. *Journal of Marine Research,* **41**, 539–56.

Sorokin, Y.I. and Mikheev, V.N. (1979). On characteristics of the Peruvian upwelling ecosystem. *Hydrobiologia,* **62**, 165–98.

Sournia, A. (1982). Form and function in marine phytoplankton. *Biological Reviews,* **57**, 347–94.

Southward, A.J. (1974). Long term changes in abundance of eggs of the Cornish pilchard (*Sardina pilchardus* Walbaum) off Plymouth. *Journal of the Marine Biological Association of the United Kingdom,* **54,** 641-9.

Southward, A.J. (1980). The western English Channel - an inconstant ecosystem? *Nature, London,* **285,** 361-6.

Spector, D.L. (ed.). (1984). *Dinoflagellates.* Academic Press, London.

Spencer, C.P. (1975). The micronutrient elements. In: *Chemical oceanography,* second edition, Volume 2. Riley, J.P. and Skirrow, G. (eds). Academic Press, London.

Steele, J.H. (1956). Plant production on the Fladen ground. *Journal of the Marine Biological Association of the United Kingdom,* **35,** 1-33.

Steele, J.H. (1957). A comparison of plant production estimates using ^{14}C and phosphate data. *Journal of the Marine Biological Association of the United Kingdom,* **36,** 233-41.

Steele, J.H. (1974). *The structure of marine ecosystems.* Blackwell Scientific Publications, Oxford.

Stewart, M.G. (1979). Absorption of dissolved organic nutrients by marine invertebrates. *Oceanography and Marine Biology Annual Review,* **17,** 163-92.

Sund, P.N., Blackburn, M. and Williams, F. (1981). Tunas and their environment in the Pacific ocean: a review. *Oceanography and Marine Biology Annual Review,* **19,** 443-512.

Sverdrup, H.U. (1953). On conditions for the vernal blooming of phytoplankton. *Journal du Conseil: Conseil permanent international pour l'exploration de la mer,* **18,** 287-95.

Sverdrup, H.U., Johnson, M.W. and Fleming, R.H. (1942). *The oceans: their physics, chemistry, and general biology.* Prentice-Hall, Inc., New Jersey.

Swift, D.G. (1980). Vitamins and phytoplankton growth. In: *The physiological ecology of phytoplankton.* Morris, I. (ed.). Blackwell Scientific Publications, Oxford.

Therriault, J.-C. and Platt, T. (1978). Spatial heterogeneity of phytoplankton biomass and related factors in the near-surface waters of an exposed coastal embayment. *Limnology and Oceanography,* **23,** 888-99.

Thornber, J.P. (1975). Chlorophyll-proteins: light harvesting and reaction center components of plants. *Annual Review of Plant Physiology,* **26,** 127-58.

Venrick, E.L. (1974). The distribution and significance of *Richelia intracellularis* Schmidt in the north Pacific central gyre. *Limnology and Oceanography,* **19,** 437-45.

Vinogradov, M.E. (1981). Ecosystems of equatorial upwellings. In: *Analysis of marine ecosystems.* Longhurst, A.R. (ed.). Academic Press, London.

Vlymen, W.J. (1970). Energy expenditure of swimming copepods. *Limnology and Oceanography,* **15,** 348-56.

Vollenweider, R.A. (ed.). (1969). *A manual on methods for measuring primary production in aquatic environments.* IBP handbook number 12. Blackwell Scientific Publications, Oxford.

Walsby, A.E. (1978). The properties and buoyancy-providing role of gas vacuoles in *Trichodesmium* Ehrenberg. *British Phycological Journal,* **13,** 103-16.

Walsh, J.J. (1981). A carbon budget for overfishing off Peru. *Nature, London,* **290,** 300-4.

Walsh, J.J., Kelley, J.C., Whitledge, T.E., MacIsaac, J.J. and Huntsman, S.A. (1974). Spin-up of the Baja California upwelling ecosystem. *Limnology and Oceanography,* **19,** 553-72.

Walsh, J.J., Whitledge, T.E., Esaias, W.E., Smith, R.L., Huntsman, S.A., Santander, H. and DeMendiola, B.R. (1980). The spawning habitat of the Peruvian anchovy, *Engraulis ringens. Deep-Sea Research,* **27,** 1-27.

White, H.H. (1979). Effects of dinoflagellate bioluminescence on the ingestion rates of herbivorous zooplankton. *Journal of experimental marine Biology and Ecology,* **36,** 217-24.

Yentsch, D.S. (1974). Some aspects of the environmental physiology of marine phytoplankton: a second look. *Oceanography and Marine Biology Annual Review,* **12,** 41-75.

Yentsch, C.S. (1980). Light attenuation and phytoplankton photosynthesis. In: *The physiological ecology of phytoplankton.* Morris, I. (ed.). Blackwell Scientific Publications, Oxford.

Yonge, C.M. (1926). Ciliary feeding mechanisms in the thecosomatous pteropods. *Journal of the Linnean Society of London. Zoology,* **36,** 417-29.

Index

Bold page numbers refer to main entries, italicized page numbers refer to illustrations.